W9-CFK-310

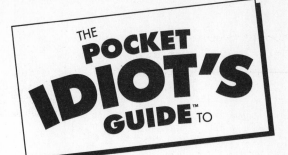
THE
POCKET
IDIOT'S
GUIDE™ TO

Algebra 1

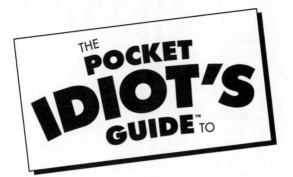

THE
POCKET
IDIOT'S
GUIDE™ TO

Algebra 1

by Denise Szecsei, Ph.D.

ALPHA

A member of Penguin Group (USA) Inc.

ALPHA BOOKS

Published by the Penguin Group

Penguin Group (USA) Inc., 375 Hudson Street, New York, New York 10014, U.S.A.

Penguin Group (Canada), 10 Alcorn Avenue, Toronto, Ontario, Canada M4V 3B2 (a division of Pearson Penguin Canada Inc.)

Penguin Books Ltd, 80 Strand, London WC2R 0RL, England

Penguin Ireland, 25 St Stephen's Green, Dublin 2, Ireland (a division of Penguin Books Ltd)

Penguin Group (Australia), 250 Camberwell Road, Camberwell, Victoria 3124, Australia (a division of Pearson Australia Group Pty Ltd)

Penguin Books India Pvt Ltd, 11 Community Centre, Panchsheel Park, New Delhi—110 017, India

Penguin Group (NZ), cnr Airborne and Rosedale Roads, Albany, Auckland 1310, New Zealand (a division of Pearson New Zealand Ltd)

Penguin Books (South Africa) (Pty) Ltd, 24 Sturdee Avenue, Rosebank, Johannesburg 2196, South Africa

Penguin Books Ltd, Registered Offices: 80 Strand, London WC2R 0RL, England

Contents

Appendixes

Introduction

Would you believe that learning algebra is exciting? Would you believe that just by buying a copy of this book your algebra skills will improve dramatically? Would you believe that buying *two* copies of this book would help you even more? Well, it is, it will, and … well, two out of three is not bad!

One of the things that separates the teens from the tweens is the study of algebra. Once you have mastered the skills necessary to perform elementary calculations (you know, adding, subtracting, multiplying, and dividing numbers), you will be ready to generalize these ideas and start thinking like a mathematician! We use the word "algebra" to refer to these generalizations.

Most algebra books are filled with symbols and variables that can bring even the sturdiest mathematics students to their knees. But as you become comfortable solving for x, factoring, multiplying, and dividing polynomials, and using the quadratic formula, your reaction to those symbols won't even register on the Richter scale. But if the symbols don't bother you, chances are that word problems will. We've saved the best for last … the last chapter will show you how to translate wordy problems into simple equations that you can then easily solve.

We've tried to put order into the operations and simplified simplifying. We've pointed out the key factors in factoring and expounded on exponents. As you read through this book, try to work out the solutions to the example problems along with us. Remember that the best way to read a math book is while holding a pencil.

Things to Help You Out Along the Way

When we talk about mathematics, it's easy to get off the subject. It's hard work for the editors, but we've insisted on including side notes in this book. You'll see the following boxes throughout this book.

Forewarning

This box alerts you to the pitfalls that unsuspecting algebraists may fall into.

Xtracts

This box contains helpful hints, useful observations, and the occasional blinding flash that often accompanies mathematics.

The Real Deal

In this box you will find definitions, formulas, and summaries of algebraic properties.

Added Information

Here you will find extra bits of information that, if nothing else, will help you when you play "Trivial Pursuit" or "Who Wants to Be a Millionaire?"

Acknowledgments

I'd like to thank Mike Sanders for giving me the opportunity to write this book, and for being patient when Hurricane Charley brought me back to basics and gave me a gentle reminder of life before technology. Jessica Faust helped coordinate things and kept communication flowing.

What you now hold in your hand was not created by me alone. Though I did type every word and symbol (sometimes twice to make them look nice!), I would not have been able to complete this project without the support of the administration and the Mathematics and Computer Science Department at Stetson University. Without their support it would be difficult to find the time to write about my favorite subject.

Kendelyn Michaels was kind enough to review the manuscript and let me know when I was being too mathematical or if I skipped too many steps along the way. Alic Szecsei was willing to help in any way he could. In addition to the extra chores he is *still* doing, his eyes were able to catch some of the mistakes and inconsistencies that I missed. Now that he's back in school, I hope to return the favor when I look over his homework!

Trademarks

All terms mentioned in this book that are known to be or are suspected of being trademarks or service marks have been appropriately capitalized. Alpha Books and Penguin Group (USA) Inc. cannot attest to the accuracy of this information. Use of a term in this book should not be regarded as affecting the validity of any trademark or service mark.

A-B-C, Easy as 1-2-3

In This Chapter

- It takes all kinds of numbers
- The amazing properties of 0 and 1
- Manipulating fractions
- The order of operations

Your first exposure to mathematics involved concrete numbers—1, 2, 3, ...—and learning how to add, subtract, multiply, and divide them. After memorizing your times tables, you branched out into fractions, decimals, and percentages. Now that you've gotten pretty good at those kinds of calculations, it's time to generalize some of their procedures.

An easy way to look at algebra is that it is a continuation and extension of the rules of arithmetic into a more general setting. It should come as no surprise, then, that the first part of learning algebra is to review the rules of arithmetic. And these rules are what this chapter is all about. Even if you think you know this stuff flat, keep reading: You might just learn something new!

Because there are rules for numbers in general, you'll need a way to represent "any number." You can't use a specific number, like 5, to represent *all* numbers. The number 5 can only speak for itself. Because of this, mathematicians and scientists, out of necessity, began using letters to represent numbers when talking in general terms. These letters are called *variables*.

The Real Deal

The word **variable** is a form of the root *vary*, which means "to make different." Therefore, any number you assign a letter can be made different.

A variable is a way to represent any old number that happens to come along. For example, you could make the letter x represent the number 5 (x would be the variable) and you could even go so far as to say that $x = 5$. Or you could say that the letter a represents the number 13,000 ($a = 13,000$). The variable is still a.

Throughout this chapter and other algebra books you will use in school, letters are used to represent numbers. And now you are going to learn about the different symbols you use with them.

Symbols Used

There are several things that you can do with numbers. You can add, subtract, multiply, and divide them. You can also compare numbers. Addition, subtraction, multiplication, and division are often called *binary operations*. A binary operation is a procedure for taking two things (hence the bi- prefix) and turning them into a third.

As you will soon see, there are really only two binary operations: addition and multiplication. Subtraction and division are just different ways of looking at addition and multiplication. But I'll talk about all four processes. In order to do that efficiently, I will make use of a variety of mathematical symbols throughout this book. It may be helpful to summarize them now, before you get too involved in the mathematics.

Symbol	Meaning	Example
+	addition	$5+2=7$
−	subtraction	$5-2=3$
· or ( )( )	multiplication	$5 \cdot 2 = 10$
/ or ÷	divided by	$\frac{5}{2} = 2\frac{1}{2}$, $5 \div 2 = 2\frac{1}{2}$
=	is equal to	$5=5$
≠	is not equal to	$5 \neq 2$
≥	is greater than	$5 \geq 2$
≥	is greater than or equal to	$5 \geq 5$, $5 \geq 2$
<	is less than	$2 < 5$
≤	is less than or equal to	$2 \leq 2$, $2 \leq 5$

While it may seem that these symbols have been used since the dawn of civilization, they are relatively new inventions. The + and − signs were introduced in 1544 and = was introduced in 1557.

Classification of Numbers

Numbers have evolved over the years, and certain sets of numbers have special names. The first numbers to be invented were the *counting*, or *natural*, *numbers*. They are the numbers 1, 2, 3, ... These numbers are also called *positive integers*. The three dots (ellipsis) mean that the list of elements is unending (infinite ∞). Notice that 0 is not in this list. That is because 0 was actually one of the last numbers to be discovered.

The *whole numbers* are the numbers 0, 1, 2, 3, ... Notice that the whole numbers are just the natural numbers together with 0.

The *negative integers* are the numbers −1, −2, −3, ...

The *integers* consist of the natural numbers (or the positive integers), their negatives, or opposites, (or the negative integers), and 0. They can be written as ..., −3, −2, −1, 0, 1, 2, 3, ...

A *rational number* is a number that can be written as the ratio of two integers. Examples of rational numbers are $\frac{1}{2}$ and $-\frac{4}{3}$. To describe a rational number in general, it helps to use variables (remember those letters?). A rational number is a number that can be written as $\frac{p}{q}$, where p and q are integers

and $q \neq 0$ (you are not allowed to divide by zero!).
Notice that $\frac{1}{2}$ and $\frac{5}{10}$ represent the same rational
number. Also, the integer 2 is a rational number,
since you can write 2 as the ratio $\frac{2}{1}$. In fact, every
integer is rational.

Xtracts

All rational numbers can be represented
by decimal numbers that either terminate or
have a nonterminating repeating pattern
(forming a group of digits that repeats
without end). For example, $\frac{2}{5} = 0.4$,
$\frac{1}{9} = 0.111...$, and $\frac{5}{27} = 0.185185185...$.

An *irrational number* is a number that cannot be written as a ratio of two integers. Irrational numbers have
a nonterminating nonrepeating decimal representa-
tion. Examples of irrational numbers are $\sqrt{2}$ and $\sqrt{5}$
(which, if you tried to write them out as a decimal,
would start out as 1.414213562373095048801... and
2.236067977499789696409... respectively).

The set of *real numbers* is formed by combining the
rational and irrational numbers.

Properties of Numbers

In addition to these classifications of numbers,
numbers can be categorized by their properties.

Prime numbers are natural numbers that can be divided evenly only by 1 and themselves. When one number *divides evenly* into another number, the remainder is 0. The numbers 3, 5, and 17 are examples of prime numbers ($\frac{3}{1} = 3$ and $\frac{3}{3} = 1$, and 3 and 1 are the *only* numbers that divide evenly into 3). *Composite numbers* are numbers that can be evenly divided by numbers other than one and themselves. The numbers 4, 9, and 12 are examples of composite numbers (since $\frac{4}{2} = 2$ with remainder 0, 2 divides into 4 evenly). All natural numbers greater than 1 can be classified as either prime or composite.

Xtracts

Remember that 1 is considered to be neither prime nor composite.

Two numbers are *relatively prime* if they share no common divisors. But relatively prime numbers are not necessarily prime numbers themselves. For example, the numbers 9 and 16 are relatively prime, since they have no common divisors ($\frac{9}{3} = 3$; $\frac{16}{3} = 5.333\ldots$). The numbers 5 and 25 are not relatively prime, since 5 divides into both of them evenly ($\frac{5}{5} = 1$; $\frac{25}{5} = 5$).

Even numbers are natural numbers that are divisible by 2 ($\frac{48}{2} = 24$). *Odd numbers* are natural numbers that are not divisible by 2 ($\frac{45}{2} = 22.5$).

Properties of Real Numbers

There are several important properties of real numbers that must be mentioned. Since I'll be talking about numbers in general, I'll need to use variables. In this discussion I will let a, b, and c denote any real number.

The first property is the idea of closure. The real numbers are *closed* under addition and multiplication. This means that if you take two *real* numbers and add them (or multiply them), what you'll end up with is a *real* number. It may be a different number, or it may be the same number, but no matter how you slice it (or add or multiply it), you'll keep getting real numbers. Case closed.

The *transitive property of equality* states that if $a = b$ and $b = c$, then $a = c$. In other words, two numbers that are both equal to a third number are equal to each other (if $x = 5$ and $5 = y$ then $x = y$).

The *trichotomy* property involves comparing two real numbers. Given two real numbers a and b, exactly one of the following is true: $a < b$, $a = b$, or $a \geq b$. So you always know where two numbers stand relative to each other ($5 < 6$; $5 = 5$; $5 > 4$).

Addition and multiplication *commute*. That is the mathematical way of saying that the order in which two numbers are added or multiplied doesn't matter. This is written symbolically as $a + b = b + a$ and $a \cdot b = b \cdot a$. For example, $3 + 7 = 7 + 3$, and $(14)(-3) = (-3)(14)$.

Addition and multiplication are also *associative*. This is a fancy way of saying that if you have a long list of numbers to add (or multiply), then you can group them in any order. This is written as $a + (b + c) = (a + b) + c$ and $a \cdot (b \cdot c) = (a \cdot b) \cdot c$. For example, $1 + (2 + 3) = (1 + 2) + 3$, and $4\left(\dfrac{1}{2} \cdot 5\right) = \left(4 \cdot \dfrac{1}{2}\right) \cdot 5$.

Forewarning

While addition and multiplication are commutative and associative, subtraction and division are not, as can be seen by the following examples:

- Subtraction doesn't commute:
  $3 - 4 \neq 4 - 3$
- Division doesn't commute:
  $4 \div 2 \neq 2 \div 4$
- Subtraction isn't associative:
  $4 - (3 - 6) \neq (4 - 3) - 6$
- Division isn't associative:
  $4 \div (5 \div 10) \neq (4 \div 5) \div 10$

While all numbers may seem to have the same importance, there are actually two numbers that stand above the rest. Those two numbers are 0 and 1. Zero is special because it is the *additive identity*. It is the unique number that you can add to any other number and have no effect. This is written as $a + 0 = a$; for example, $5 + 0 = 5$. It may not seem like such a big deal to you ... after all, nothing is nothing. But the notion of nothing in mathematics took a long time to be quantified.

The number 1 is special for a similar reason; 1 is the *multiplicative identity*. It is the *only* number that you can multiply any other number by and have no effect. This idea is illustrated by the equation $a \cdot 1 = a$; for example, $5 \cdot 1 = 5$. These two numbers will play a role in most of the algebraic problems that you face.

One of the ways that the identities come into play in algebra is in the development of the inverse. For every real number a, there is a unique real number, called the *additive inverse* of a, and denoted $-a$, such that $a + (-a) = 0$. In other words, the additive inverse of a is the unique number that you add to a in order to get 0 (the additive identity). Every real number has an additive inverse. Notice that 0 is its own additive inverse. Subtraction is then defined in terms of addition (which is why subtraction is not considered a binary operation in its own right): $a - b = a + (-b)$.

You may have heard that a negative times a negative is a positive. You may have been told to just memorize it, use it, and trust that it works. Now would be a good time to talk about why that is true. When you write $-a$, you are writing the additive inverse of a: the unique number that, when it is added to a, gives you 0. In other words, $a + (-a) = 0$. What would be the additive inverse of $-a$? Well, in keeping with our notation, it would be $-(-a)$. But wait a minute! Can't you add a to $-a$ and get 0? Doesn't $(-a) + a = 0$? So now you have two additive inverses for $-a$: a and $-(-a)$. You can't have two distinct additive inverses (additive inverses are unique, which means that each number a can only have one). The only option is that a and $-(-a)$ are the same thing: $-(-a) = a$. Sometimes that's read that a negative times a negative equals a positive. Mystery solved!

The Real Deal _____

There are several versions of this rule. One version is, "A negative times a negative is a positive." Another way to look at it is, "The opposite of the opposite is the original: $-(-a) = a$."

It's time to turn our attention to multiplicative inverses. For each real number a, except 0, there is a unique real number, called the multiplicative inverse and denoted a^{-1}, such that $a \cdot a^{-1} = 1$. Just as with the additive inverses, the multiplicative inverse of a is the unique number that you multiply a by in order to get 1 (the multiplicative inverse). Notice that 0 (the additive inverse) is the only real number that doesn't have a multiplicative inverse. The multiplicative inverse of a number is also called the *reciprocal* of that number. Sometimes the multiplicative inverse of a is written as $\frac{1}{a}$. Division is then defined in terms of multiplication. If $b \neq 0$, then

$$a \div b = \frac{a}{b} = a\left(\frac{1}{b}\right) = a\left(b^{-1}\right).$$

The Real Deal

The reciprocal (or inverse) of a is denoted a^{-1}, and $a^{-1} = \dfrac{1}{a}$.

The last property of the real numbers that I will discuss has to do with how they behave when you combine addition and multiplication. It is called the *distributive* property. The property can be stated symbolically as: $a \cdot (b + c) = a \cdot b + a \cdot c$.

You can think of a being distributed to both b and c. This property can be thought of as "multiplication distributes over addition."

There are several properties of real numbers that will be used throughout this book. I've summarized some of them here, and have included some examples to help illustrate the ideas involved.

Property	*Example*
$a - b = a + (-b)$	$5 - 8 = 5 + (-8) = -3$
$a - (-b) = a + b$	$5 - (-8) = 5 + 8 = 13$
$-a = -1 \cdot a$	$-3 = (-1)(3)$
$a \cdot (b - c) = a \cdot b - a \cdot c$	$3(8 - 3) = 3 \cdot 8 - 3 \cdot 3$ $= 24 - 9 = 15$
$-(a + b) = -a - b$	$-(2 + 4) = -2 - 4 = -6$
$-(a - b) = -a + b$	$-(7 - 4) = -7 + 4 = -3$
$(-a)(-b) = a \cdot b$	$(-3)(-4) = 3 \cdot 4 = 12$
$0 \cdot a = 0$	$0 \cdot 5 = 0$

$$\frac{a}{1} = a \qquad\qquad \frac{6}{1} = 6$$

$$\frac{0}{a} = 0 \quad (a \neq 0) \qquad\qquad \frac{0}{4} = 0$$

$$\frac{a}{a} = 1 \quad (a \neq 0) \qquad\qquad \frac{5}{5} = 1$$

$$a\left(\frac{b}{a}\right) = b \quad (a \neq 0) \qquad\qquad 2\left(\frac{5}{2}\right) = 5$$

Properties of Zero

Zero is an amazing number as well as an interesting idea. How does one put a name to nothing? There are several properties that only belong to 0.

The first two properties of 0 have already been observed. Zero is the only number that is its own additive inverse: $0 + 0 = 0$. Zero is also the only real number that does not have a multiplicative inverse. In other words, there is no real number that you can multiply 0 by and get 1. The reason for this is that 0 times any real number is 0. You can't turn 0 into 1 through multiplication by a real number.

Another special property of 0 is the following: If two real numbers a and b are multiplied together and the result is 0, then either $a = 0$ or $b = 0$. No other real number has this property. Try it with, say, 2. If $a \cdot b = 2$, then what can you conclude about a and b? Absolutely nothing … well, actually you can conclude that neither a nor b are 0! But that's about it. There are quite a few pairs of numbers that, when multiplied, give 2. For example, $\frac{1}{3} \cdot 6 = 2$, and $\frac{2}{5} \cdot 5 = 2$. Those are just two pairs of numbers whose product is 2. There are many more where those came from! So there's not a lot that you can conclude about a and b, if all you know is that their product is 2. But 0 is a whole other story. If $a \cdot b = 0$, then you know beyond a shadow of a doubt that either $a = 0$ or $b = 0$. And 0 is the only real number with that property!

Absolute Value

The absolute value of a real number a, denoted $|a|$, is the magnitude of the number. It gives a measure of how large that number is. The absolute value of a number can also be thought of as the distance that the number is from 0. For example, $|3| = 3$ and $|-3| = 3$ because both 3 and -3 are 3 units from 0. Notice that $|0| = 0$.

It's time to make some observations about the absolute value of a real number. First of all, 0 is the only number whose absolute value is 0. The second observation is that the absolute value of a number can never be negative. If a is any real number, then $|a| \geq 0$. The final property that I want to point out is that if a is any real number, then $|a| = |-a|$.

You can use the properties of real numbers to come up with a formula for the absolute value of a number. Since $-(-a) = a$, you can write the absolute value of any real number a as:

$$|a| = \begin{cases} a & a \geq 0 \\ -a & a < 0 \end{cases}$$

Check to see that this formula works. To find $|3|$, you would use the top rule, since $3 \geq 0$. The formula says that $|3| = 3$. To find $|-3|$, you would use the bottom rule, since $-3 < 0$. The formula gives $|-3| = -(-3) = 3$.

Manipulating Rational Numbers

Remember that rational numbers are numbers that can be written as the ratio of two integers. Rational numbers are also called *simple* fractions. If a number is written in the form $\frac{a}{b}$, it means $a \div b$. The number a is called the numerator and b is called the denominator. If a and b are both positive numbers, then $\frac{a}{b}$ is called a *proper* fraction if $a < b$, an improper fraction if $a > b$, and a whole number if b divides evenly into a. This is a good time to review the basic operations involving fractions.

Multiplying Fractions

Multiplication of fractions is fairly straightforward. When multiplying fractions, simply multiply the numerators of the fractions together to get the new numerator, and multiply the denominators of the fractions together to get the new denominator.

Example 1: Multiply the fractions $\frac{2}{3}$ and $\frac{5}{13}$.

Solution: $\frac{2}{3} \cdot \frac{5}{13} = \frac{10}{39}$.

Reducing Fractions

If the greatest common factor of the numerator and denominator of a given fraction is 1, then you say that the fraction is in *lowest terms*, or is *reduced*. If the greatest common factor is not 1, then you can put the fraction into reduced form by dividing both the numerator and denominator by this greatest common factor. This is often referred to as canceling.

Example 2: Reduce the fraction $\dfrac{28}{64}$.

Solution: Factor the numerator and denominator, look for common factors and cancel them:

$$\frac{28}{64} = \frac{\cancel{4} \cdot 7}{\cancel{4} \cdot 16} = \frac{7}{16}$$

The reason you can do this has to do with the properties of 1. What you are really doing is breaking the fraction up into pieces and then making use of the properties of the multiplicative identity:

$$\frac{28}{64} = \frac{4 \cdot 7}{4 \cdot 16} = \frac{4}{4} \cdot \frac{7}{16} = 1 \cdot \frac{7}{16} = \frac{7}{16}$$

Multiplying Fractions (Again)

One problem with multiplying fractions is that the numbers can get pretty big. It's harder to factor big numbers than it is to factor small ones. There are ways to keep the numbers small; if any of the terms in the numerator and denominator share a common factor, it is easiest to cancel the common factor and then proceed with the multiplication.

Example 3: Multiply the fractions $\frac{28}{39}$ and $\frac{13}{16}$.

Solution: One way is to multiply numerators and denominators and then reduce:

$$\frac{28}{39} \cdot \frac{13}{16} = \frac{364}{624} = \frac{\cancel{52} \cdot 7}{\cancel{52} \cdot 12} = \frac{7}{12}$$

Notice that you have to find the factors of 364 and of 624. Factoring them will take time. Alternatively, if the numerators and denominators are factored, and common factors canceled, then the multiplication is much easier:

$$\frac{28}{39} \times \frac{13}{16} = \frac{\cancel{2} \cdot \cancel{2} \cdot 7 \cdot \cancel{13}}{3 \cdot \cancel{13} \cdot \cancel{2} \cdot \cancel{2} \cdot 2 \cdot 2} = \frac{7}{3 \cdot 2 \cdot 2} = \frac{7}{12}$$

Dividing Fractions

A *complex* fraction is a fraction where the numerator, denominator, or both contain a fraction. A complex fraction can be converted into a simple fraction using a simple rule: Invert the fraction in the denominator and then multiply the numerator and the inverted denominator.

Example 4: Convert the complex fraction $\frac{2/7}{5/9}$ into a simple fraction.

Solution: Invert the denominator and multiply:

$$\frac{2/7}{5/9} = \frac{2}{7} \cdot \frac{9}{5} = \frac{18}{35}$$

Adding and Subtracting Fractions, Part I

Adding and subtracting fractions is a little more complicated than multiplying or dividing them. In order to add or subtract fractions, the denominators of the fractions must be the same. If the denominators are not the same, then you have to use your properties of real numbers to make them the same. The main property to make use of is the multiplicative identity.

Example 5: Add the fractions $\frac{4}{9}$ and $\frac{1}{5}$.

Solution: You first have to find the common denominator. Because it is easier to work with small numbers, try to find the smallest number that is evenly divisible by 9 and by 5. Since 9 and 5 have no common factors (they are relatively prime), the common denominator in this case would be 45. The only number you can multiply each fraction by without changing it is 1. You will actually multiply each fraction by a disguised form of 1 (the multiplicative identity). The fraction $\frac{4}{9}$ will be multiplied by $\frac{5}{5}$, and the fraction $\frac{1}{5}$ will be multiplied by = . Then you can go ahead and add the fractions. Be sure to put your answer into a pretty form (reduced, if possible):

$$\frac{4}{9} + \frac{1}{5} = \frac{4}{9} \cdot \frac{5}{5} + \frac{1}{5} \cdot \frac{9}{9} = \frac{20}{45} + \frac{9}{45} = \frac{29}{45}$$

Be mindful when working out these types of problems. It is easy to get into the habit of canceling common divisors, but in this problem you are actually introducing common divisors into the mix. You need to have the denominators of the two fractions be the same. Canceling common factors in the middle of the problem will just unravel the tangled web that you are trying to weave!

Order of Operations

Order matters. Just as you have to look *before* you leap, and you have to eat your vegetables *before* you get your dessert, you have to multiply *before* you add. If a mathematical expression involves more than one binary operation, things can get awkward. The rules for deciding which operation takes precedence must be established in order to avoid confusion. In the grand scheme of things, multiplication and division (read left to right) comes first, followed by addition and subtraction. Of course, if you want things to be done differently, you can use parentheses to alter the order. Whatever happens inside parentheses is done first: Start with the innermost parentheses and work your way out.

It is important to be aware of the order of operations when evaluating expressions in a calculator. Calculators that abide by the standard order of operations are sometimes called "smart" calculators. There are calculators that do not know the order of operations, and care must be taken when using such calculators. For example, if you enter the expression $7 + 3 \cdot 5$ into a smart calculator, you should get 22. If the proper order of operations is not followed, then your answer would be 50. Now would be a good time to determine whether your calculator is smart or not; go ahead, punch in $7 + 3 \cdot 5$.

Adding and Subtracting Fractions, Part II

When adding or subtracting fractions, it is important to pay attention to the order of operations. When dealing with fractions, sometimes parentheses are assumed and not written. You must learn to read between the lines, and perform the operations in the right order.

Example 6: Simplify the fraction $\dfrac{14+20}{21+15}$.

Solution: You must be careful when you see fractions written like this. The parentheses have been omitted by convention, but you have to act as if they are there. You should treat this fraction as $\dfrac{(14+20)}{(21+15)}$, and follow the instructions in parentheses first, and then do the division (and cancel where possible, *after* you've added):

$$\frac{(14+20)}{(21+15)} = \frac{34}{36} = \frac{\cancel{2}\cdot 17}{\cancel{2}\cdot 2\cdot 3\cdot 3} = \frac{17}{18}$$

Forewarning

Don't even think about doing any cancellation before dealing with the addition in the numerator and denominator. Canceling before adding (or subtracting) is a major (and common) mistake!

Curses! FOILed Again!

Parentheses can be used to boost addition over multiplication. In order to evaluate the expression $(2 + 4)(5 + 3)$, you must first add the 2 and the 4, then add the 5 and the 3, and then, finally, multiply the two results:

$$(2 + 4)(5 + 3) = (6)(8) = 48$$

But you can also use the distributive property (twice!) to expand the expression and get rid of the parentheses. This makes things look more complicated, as you will see:

$$(2 + 4)(5 + 3) = 2 \cdot (5 + 3) + 4 \cdot (5 + 3)$$
$$= 2 \cdot 5 + 2 \cdot 3 + 4 \cdot 5 + 4 \cdot 3$$
$$= 10 + 6 + 20 + 12$$
$$= 48$$

Comparing these two methods of solving the same problem, you may think that the first way—using the order of operations—is easier. When you are working with specific real numbers, I would agree with you. You may think that mathematicians look for ways to make solving problems more difficult, but trust me when I tell you that you will use this application of the distributive property over and over again when you simplify algebraic expressions.

Because addition and multiplication commute, the order that the expansion is written in doesn't matter. Just make sure that you include all of the pieces. There is an easy way to visualize this expansion method when you multiply $(a + b) \cdot (c + d)$. This method is often referred to as the FOIL method: **F**irst terms (a and c), **O**utside terms (a and d), **I**nside terms (b and c), **L**ast terms (b and d). You can visualize a face (of sorts) when you use this method, as shown in Figure 1.1. Making a face is one way to make sure that you include all necessary combinations of terms.

$$(a + b)(c + d) = a \times c + a \times d + b \times c + b \times d$$

The face of FOIL.

Example 7: Expand the expression $(3+2)(6+4)$ using the FOIL method.

Solution: $(3 + 2)(6 + 4) = 3 \cdot 6 + 3 \cdot 4 + 2 \cdot 6 + 2 \cdot 4$
$= 18 + 12 + 12 + 8 = 50$.

Evaluating Expressions

An expression is just a statement that combines numbers and variables in a meaningful way using mathematical operations. It's kind of like a fragmented phrase in English. For example, the expression $a + 4$ is an expression that means take the number a and add 4 to it. An expression can involve any number of variables and letters. Another example of an expression is $\frac{1}{2}(a + b)$. This expression represents how to find the average of two real numbers a and b. Remember that dividing by 2 and multiplying by $\frac{1}{2}$ are the same thing.

To evaluate expressions for particular values of the variables involved, just substitute in the values for the variables and perform the operations in the proper order.

Example 8: Evaluate the expression $\frac{1}{2}(a+b)$ when $a = 4$ and $b = 10$.

Solution: Start with the expression $\frac{1}{2}(a+b)$, and replace a with 4 and b with 10: $\frac{1}{2}(4+10)$. Add the two numbers inside the parentheses (following our order of operations): $\frac{1}{2} \cdot (14)$. Finish up the problem by multiplying $\frac{1}{2}$ and 14 to get 7; so when $a = 4$ and $b = 10$, the expression $\frac{1}{2}(a+b)$ is 7.

The Least You Need to Know

- The real numbers have several wonderful properties under addition and multiplication: they're closed, they commute, they associate, and multiplication distributes over addition.

- 0 is the additive identity, and the additive inverse of a number is the number you have to add to make the sum 0.

- 1 is the multiplicative identity, and the multiplicative inverse (or reciprocal) of a number is the number you have to multiply by to make the product 1.

- Any number times 0 is 0, and 0 is the only number with the property that if $a \cdot b = 0$, then either $a = 0$ or $b = 0$.

- The law of trichotomy is this: Given any two real numbers a and b, either $a > b$, $a = b$, or $a < b$.

You've Got the Power

In This Chapter

- Products and quotients of powers
- Powers of products and quotients
- The power of zero
- Squaring sums

Numbers can be very big. For example, a googol is the number 1 followed by 100 zeros. A googolplex is the number 1 followed by a googol of zeros. It would take more than my lifetime to write all of the zeros involved in a googolplex.

Fortunately, mathematicians have developed a shorthand notation for large numbers. This notation involves using exponents, or powers. After reading this chapter you will be able to write numbers even bigger than a googolplex in no time at all!

Powers

Exponents can be used to shorten some expressions involving repeated multiplication. For example, the product of $2 \cdot 2 \cdot 2 \cdot 2 \cdot 2 \cdot 2$ involves multiplying 2 by itself 6 times. Exponential expressions enable us to represent this idea without having to type out a bunch of 2s. The rules you are about to see only hold true when dealing with nonzero numbers!

Positive Integer Powers

The product of $2 \cdot 2 \cdot 2 \cdot 2 \cdot 2 \cdot 2$ can be abbreviated 2^6 (read "two to the sixth"). The number 6 in this expression is called the *exponent*, or the *power*, and the number 2 in this expression is called the *base*.

To expand an exponential expression, the base and the exponent need to be identified. For example, in the expression 3^5, the base is 3 and the exponent is 5. In the expression -4^3, the base is 4 and the exponent is 3; even though the parentheses have been omitted, you are still expected to know where they belong: $-4^3 = -(4^3)$. The negative sign in the beginning means that the number is less than 0. There is a difference between -2^4 and (-2^4):

$$-2^4 = -(2 \cdot 2 \cdot 2 \cdot 2) = -16$$

$$(-2)^4 = (-2) \cdot (-2) \cdot (-2) \cdot (-2) = 16$$

In the first expression, the base is 2, and in the second expression the base is –2. What a difference parentheses make!

Added Information

A googol is 10^{100} (1 followed by 100 zeros) and a googolplex is 10^{googol} (1 followed by a googol of zeros)!

In general, an exponential expression is an expression of the form a^n, where a is the base, and n is the exponent. At this point, the expression only makes sense if n is a positive integer. The expression a^n is read "a to the nth power" or "the nth power of a" or "a to the n." Special powers are a^2 (read "a squared") and a^3 (read "a cubed").

Example 1: Write $5 \cdot 5 \cdot 5 \cdot 5 \cdot 5 \cdot 5 \cdot 5 \cdot 5 \cdot 5$ as an exponent.

Solution: Since 5 appears in the expression 9 times, $5 \cdot 5 \cdot 5 \cdot 5 \cdot 5 \cdot 5 \cdot 5 \cdot 5 \cdot 5 = 5^9$.

Negative Integer Powers

What happens if the powers are negative integers? Recall the notation for the multiplicative inverse (or reciprocal) of a real number a: $a^{-1} = \dfrac{1}{a}$. Using that relationship, you can interpret exponential expressions when the powers are negative integers. Consider the exponential expression 2^{-5}. This can be thought of as $(2^{-1})^5$. In other words, the base is the reciprocal of 2 (i.e., $\frac{1}{2}$), and the exponent is 5. So $\left(2^{-1}\right)^5 = \dfrac{1}{2} \cdot \dfrac{1}{2} \cdot \dfrac{1}{2} \cdot \dfrac{1}{2} \cdot \dfrac{1}{2}$.

In general, $a^{-n} = \left(\dfrac{1}{a}\right)^n = \dfrac{1}{a^n}$, where n is a positive integer. The negative power just means that the base goes into the denominator!

Example 2: Write 4^{-3} without exponents.

Solution: $4^{-3} = \left(\dfrac{1}{4}\right)^3 = \dfrac{1}{4^3} = \dfrac{1}{4 \cdot 4 \cdot 4} = \dfrac{1}{64}$.

Power Rules

When you multiply or divide exponential expressions, there are some rules that can help simplify the calculation. Let's look at an example and generalize what you observe.

Suppose that you need to multiply 2^3 and 2^4. If you were to write each piece out, you would have this:

$$2^3 \cdot 2^4 = (2 \cdot 2 \cdot 2) \cdot (2 \cdot 2 \cdot 2 \cdot 2) =$$
$$2 \cdot 2 \cdot 2 \cdot 2 \cdot 2 \cdot 2 \cdot 2 = 2^7$$

Notice that if you add the powers of the 2s on the left, you end up with the power of 2 given on the right! That observation can be generalized to the following:

$$a^n \cdot a^m = a^{n+m}$$

Basically, if you multiply two numbers with the same base, you add the exponents.

Let's look at what happens with division. Suppose you want to divide 2^5 by 2^3. If you expand the powers in the numerator and denominator, you will see the opportunity for cancellation:

$$\frac{2^5}{2^3} = \frac{2 \cdot 2 \cdot \cancel{2} \cdot \cancel{2} \cdot \cancel{2}}{\cancel{2} \cdot \cancel{2} \cdot \cancel{2}} = 2 \cdot 2 = 2^2$$

Notice that if you subtract the power of 2 in the denominator from the power of 2 in the numerator on the left side of the equation above, you end up with the power of 2 on the right side of the equation. This fact can also be generalized:

$$\frac{a^n}{a^m} = a^{n-m}$$

Basically, this rule says that if you divide two numbers with the same base, you subtract the exponent of the denominator from the exponent of the numerator.

Things *can* (and will) get a bit more complicated. For example, what if you have the expression $(2^3)^4$? Well, if you carefully write out what this expression means, and then count the number of 2s involved, the pattern will be revealed. You can write this:

$$(2^3)^4 = (2^3)(2^3)(2^3)(2^3) = 2 \cdot 2 \cdot 2 \cdot 2 \cdot 2 \cdot 2 \cdot 2 \cdot 2 \cdot 2 \cdot 2 \cdot 2 \cdot 2 = 2^{12}$$

If you take the product of the powers on the left side of the equation, notice that you get the power on the right side of the equation. This observation generalizes to this:

$$(a^n)^m = a^{n \cdot m}$$

Basically, if you raise a power to a power, you multiply the two powers.

The Real Deal

The product of powers adds exponents: $a^n \cdot a^m = a^{n+m}$. The quotient of powers subtracts exponents: $\dfrac{a^n}{a^m} = a^{n-m}$.
The power of powers multiplies exponents: $(a^n)^m = a^{n \cdot m}$.

There are a couple of other observations worth mentioning. They can be easily verified by an example or two:

$$\frac{1}{a^{-n}} = a^n \qquad \frac{a^n}{a^m} = \frac{1}{a^{m-n}}$$

Of course, these rules only hold when the bases of the exponential expressions are the same. If the bases are different, then you have to expand each exponential expression and try to simplify it using the properties of real numbers.

The amazing thing about these rules for exponents is that it doesn't matter whether the *exponents* are positive or negative.

Familiarizing yourself with the rules for manipulating exponential expressions will help you in your algebraic endeavors later on in this book.

What About Zero?

The rules for exponents were developed with positive integers in mind. They were expanded to incorporate the negative integers, but you may be wondering how to deal with an exponent of 0.

You can explore this situation using the rule for dividing exponents and remembering the properties of 1. If, in general, $\frac{a^n}{a^m} = a^{n-m}$, let's see what happens when n and m are equal to each other. Suppose, for simplicity, that both m and n are 1. Then $\frac{a^1}{a^1} = a^{1-1} = a^0$. But $\frac{a^1}{a^1} = 1$. Now you can make sense of the case where the exponent is 0: $a^0 = 1$. Of course, this only holds if $a \neq 0$.

The Real Deal _____

Don't forget that $a^0 = 1$ if $a \neq 0$; 0^0 and 0^{-n} if $n > 0$ are not defined.

Powers of Quotients and Products

There are times when you have a product of numbers that you are raising to a power. For example, you may have to evaluate $(2 \cdot 3)^3$. In algebra, everything is fair. Both the 2 and the 3 appear in the base, and both of them will be cubed: $(2 \cdot 3)^3 = 2^3 \cdot 3^3$. One of the things to keep in mind is that algebra, if nothing else, is fair. You can generalize this into a product power rule: $(a \cdot b)^n = a^n \cdot b^n$.

Exponentiating quotients (raising a quotient to a power) occurs in much the same way. If you raise a quotient to a power, you raise both the numerator and denominator to that same power. For example, $\left(\frac{3}{5}\right)^4 = \frac{3^4}{5^4}$. This, too, will generalize into a quotient power rule: $\left(\frac{a}{b}\right)^n = \frac{a^n}{b^n}$.

The Real Deal _____

The power of a product: $(a \cdot b)^n = a^n \cdot b^n$.

The power of a quotient: $\left(\frac{a}{b}\right)^n = \frac{a^n}{b^n}$.

Powers with Addition or Subtraction

Exponentiation distributes over multiplication and division: $(a \cdot b)^n = a^n \cdot b^n$ and $\left(\frac{a}{b}\right)^n = \frac{a^n}{b^n}$. When powers and addition or subtraction are combined, things can get a bit complicated. It's worth spending some time exploring this with numbers now, because you'll come across it with variables throughout your study of algebra.

Let's work out a problem involving a power of 2 and addition first.

Example 2: Evaluate $(3 + 6)^2$ two ways: first by using the order of operations, and second by using FOIL.

Solution: Using the order of operations: $(3 + 6)^2 = 9^2 = 81$.

Using FOIL:

$$(3 + 6)^2 = (3 + 6)(3 + 6)$$
$$= 3 \cdot 3 + 3 \cdot 6 + 6 \cdot 3 + 6 \cdot 6$$
$$= 9 + 18 + 18 + 36$$
$$= 81$$

I'll be the first to admit that the order of operations makes the evaluation proceed quickly, but if you practice using FOIL with numbers (where you have a way to check your work), then when you have variables instead of numbers you will already be comfortable with the method.

Notice that when you evaluated $(3 + 6)^2$ using FOIL, there were 4 terms: $3 \cdot 3$, $3 \cdot 6$, $6 \cdot 3$, and $6 \cdot 6$. The first and last terms are just the squares of each term in parentheses (3^2 and 6^2). The middle terms, $6 \cdot 3$ and $3 \cdot 6$, are the same product (thanks to the commutative property of multiplication). You can write $6 \cdot 3 + 3 \cdot 6$ as $2 \cdot (3 \cdot 6)$. So the expansion of $(3 + 6)^2$ using FOIL gives $(3 + 6)^2 = 3^2 + 2 \cdot (3 \cdot 6) + 6^2$.

The first term is the square of the first term in parentheses, the last term is the square of the second term in parentheses, and the middle term is twice the product of the first and second terms in parentheses. Let's practice that again.

The Real Deal

The square of a sum (and difference) appears throughout algebra and is well worth memorizing: $(a + b)^2 = a^2 + 2(a \cdot b) + b^2$ and $(a - b)^2 = a^2 - 2(a \cdot b) + b^2$. When using FOIL to expand the square of a sum, there are *three* terms involved. Remember to include that middle term: twice the product of the two terms.

Example 3: Expand $(2 + 8)^2$ using FOIL.

Solution: $(2 + 8)^2 = 2^2 + 2 \cdot (2 \cdot 8) + 8^2 = 4 + 32 + 64 = 100$.

Let's try an example that involves subtraction.

Example 4: Expand $(6 - 2)^2$ two ways: first by using the order of operations, and second by using FOIL.

Solution: First, using the order of operations: $(6 - 2)^2 = 4^2 = 16$. Second, using FOIL (and making use of the observation above):

$$(6 - 2)^2 = (6 + (-2))^2$$

$$= 6^2 + 2 \cdot (6 \cdot (-2)) + (-2)^2$$

$$= 36 - 24 + 4$$

$$= 16$$

Same answer, different methods. Notice that $(6 - 2)^2 = 6^2 - 2 \cdot (6 \cdot 2) + 2^2$. The only effect of the negative sign is that instead of adding the middle term, you subtract it! Before you move on, let's summarize the similarities and differences between squaring the sum of terms and squaring the difference between terms. If a and b are numbers, then $(a + b)^2 = a^2 + 2(a \cdot b) + b^2$ and $(a - b)^2 = a^2 - 2(a \cdot b) + b^2$.

Cubing things is a bit more complicated. FOIL works great for squares, but for cubes it's best to wait until you get knee deep into algebra and can learn about Pascal's triangle.

Forewarning

Remember that $(a + b)^n$ is *different* than $a^n + b^n$; $(a + b)^n \neq a^n + b^n$. For example: $(1 + 2)^4 = 3^4 = 81$, whereas $1^4 + 2^4 = 1 + 16 = 17$. The second expression is missing all of the mixed (or middle) terms.

The Least You Need to Know

- The power rules are as such: $a^n \cdot a^m = a^{n+m}$, $\dfrac{a^n}{a^m} = a^{n-m}$, and $(a^n)^m = a^{n \cdot m}$

- If $a \neq 0$, then $a^0 = 1$.

- Powers of products and quotients: $(a \cdot b)^n = a^n \cdot b^n$ and $\left(\dfrac{a}{b}\right)^n = \dfrac{a^n}{b^n}$.

- FOILing squares of sums and differences: $(a + b)^2 = a^2 + 2(a \cdot b) + b^2$ and $(a - b)^2 = a^2 - 2(a \cdot b) + b^2$.

Finding Your Roots

In This Chapter

- Simplifying radicals
- Roots as fractional exponents
- Roots of products and quotients
- Rationalizing the denominator

Roots and powers are interconnected. You are probably familiar with square roots of numbers. For example, the square root of 9 is 3, because $3^2 = 9$ (3 squared is 9). It works the same for other roots: since $3^3 = 27$ (3 cubed is 27), then the cube root of 27 is 3. If $a^n = b$, then the nth root of b is a.

In this chapter, you will learn about exponents from the underside; you will get to examine their roots.

Your Real Roots

Taking roots undoes raising powers. The expression $\sqrt[n]{a}$ is read as the "nth root of a." The root sign $\sqrt{}$ is called a *radical*, the expression under the radical sign is called the *radicand*, and the root n is called the *index*.

Added Information

The radical sign was introduced in 1525—almost 20 years *before* the introduction of + and − signs!

A given number may have a single real root, two real roots, or no real roots. For example, 27 only has one cube root, because 3 is the only number whose cube is 27. The number 4 has two real square roots, because there are two numbers that, when squared, yield 4: $2^2 = 4$ and $(-2)^2 = 4$. The number −1 has no real square roots: I dare you to try and find a real number that, when multiplied by itself, gives −1. To understand this, think about the real numbers. There are three kinds of real numbers: positive real numbers, negative real numbers, and 0. If you take any positive number and square it, you'll get a positive number. If you take 0 and square it, you'll get 0. If you take any negative number and square it, you'll get a positive number. Those are the only kinds of real numbers there are, and none of them, when squared, can give −1. In fact, there isn't a single negative number that has a real square root. That's why most of the properties of roots (and tricks for evaluating them) only apply for positive real numbers.

Forewarning

Dealing with negative numbers requires extra thought. Proceed cautiously when dealing with even roots of negative numbers.

If a number has two real nth roots, one is always the negative of the other. The positive root is called the *principal root*. Use the radical sign $\sqrt{\ }$ for the principal root of a real number. If there is no number in the upper left corner of the radical, then it's a square root. A cube root would be written $\sqrt[3]{\ }$, and the nth principal root would be written $\sqrt[n]{\ }$. In our previous discussion, you could have saved some ink by writing $\sqrt{9} = 3$, $\sqrt[3]{27} = 3$, and $\sqrt{4} = 2$.

Simplifying Your Roots

A square root expression is considered *simplified* if the radical has no repeating factors. It may be helpful if you use the rule $\sqrt{a^2b} = a\sqrt{b}$. The best way to simplify a square root is to factor the radicand completely, and pick out the multiple factors.

Forewarning

Remember to use the simplification $\sqrt{a^2b} = a\sqrt{b}$ wisely. Taking even roots of negative numbers must be done carefully.

Example 1: Simplify $\sqrt{120}$.

Solution: Factor 120 into its prime factors: $120 = 2 \cdot 2 \cdot 2 \cdot 5 \cdot 3$. Since the factor 2 shows up 3 times, you can pull one of pairs of 2s out from under the radical:

$$\sqrt{120} = \sqrt{2 \cdot 2 \cdot 2 \cdot 5 \cdot 3} = 2\sqrt{2 \cdot 5 \cdot 3} = 2\sqrt{30}$$

To simplify higher-powered radicals, you still need to factor the radicand. The rule that is the most helpful is a generalization of the square root property: $\sqrt[n]{a^n b} = a\sqrt[n]{b}$. When n is even and the radicand involves negative numbers, proceed with caution.

Rationalizing Your Roots

Now that you understand roots, it's time to put them in terms of exponents. If a is a real number, and n is a positive integer, you can write $\sqrt[n]{a} = a^{\frac{1}{n}}$. You can look at roots as just being fractional (or rational) exponents. If a is a real number and m and n are integers, then $a^{\frac{n}{m}} = \left(a^{\frac{1}{m}}\right)^n = \left(\sqrt[m]{a}\right)^n$ (assuming that $\sqrt[m]{a}$ exists). You can also change the order using the properties of exponents: $a^{\frac{n}{m}} = \left(a^n\right)^{\frac{1}{m}} = \sqrt[m]{a^n}$.

Example 2: Simplify the following expressions:

a) $4^{\frac{3}{2}}$

b) $(27)^{-\frac{1}{3}}$

c) $(-8)^{\frac{5}{3}}$

Solution:

a) $4^{\frac{3}{2}} = \left(4^{\frac{1}{2}}\right)^3 = \left(\sqrt{4}\right)^3 = 2^3 = 8$; or

$\quad 4^{\frac{3}{2}} = \left(4^3\right)^{\frac{1}{2}} = 64^{\frac{1}{2}} = \sqrt{64} = 8$

b) $(27)^{-\frac{1}{3}} = \left(27^{\frac{1}{3}}\right)^{-1} = \left(\sqrt[3]{27}\right)^{-1} = (3)^{-1} = \frac{1}{3}$

c) $(-8)^{-\frac{5}{3}} = \left((-8)^{\frac{1}{3}}\right)^5 = \left(\sqrt[3]{-8}\right)^5 = (-2)^5 = -32$

Root Rules (or Roots Rule)

Roots have the same multiplication and division properties that powers do. There are product and quotient root rules, as well as a root of a root rule (try saying that five times fast!).

The first root rules deal with combining roots with the same radicand. Recall that when you multiply two exponents with the same base, you add the exponents. Now that you see roots in their true light (they just represent fractional exponents), in order to multiply two roots with the same radicand, you must put things in exponential form:

$$\sqrt[n]{a} \cdot \sqrt[m]{a} = a^{\frac{1}{n}} \cdot a^{\frac{1}{m}} = a^{\frac{1}{n}+\frac{1}{m}} = a^{\frac{m+n}{mn}} = \sqrt[mn]{a^{m+n}}$$

This root rule is hardly worth memorizing, since it comes from the rules for multiplying two exponentials with the same base (which *is* worth memorizing!). The twist is that roots involve fractional exponents, so multiplying roots involves adding fractions: getting a common denominator, etc.

It should then come as no surprise that division of roots with the same radicand involves subtracting fractions:

$$\frac{\sqrt[n]{a}}{\sqrt[m]{a}} = \frac{a^{\frac{1}{n}}}{a^{\frac{1}{m}}} = a^{\frac{1}{n}-\frac{1}{m}} = a^{\frac{m-n}{mn}} = \sqrt[mn]{a^{m-n}}$$

When you raise a power to a power, you multiply the powers. Since roots and powers are related to each other, this property carries over into the roots as well. When you take the root of a root, you multiply the roots: $\sqrt[m]{\sqrt[n]{a}} = \sqrt[mn]{a}$.

Let's work out a couple of examples to illustrate these rules.

Example 3: Simplify the following exponential expressions:

a) $\sqrt[4]{3} \cdot \sqrt[3]{3}$

b) $\dfrac{\sqrt{5}}{\sqrt[3]{5}}$

c) $\sqrt[3]{\sqrt[5]{10}}$

Solution: Each of the expressions is expanded and simplified:

a) $\sqrt[4]{3} \cdot \sqrt[3]{3} = 3^{\frac{1}{4}} \cdot 3^{\frac{1}{3}} = 3^{\frac{1}{4}+\frac{1}{3}} = 3^{\frac{3}{12}+\frac{4}{12}} = 3^{\frac{7}{12}} = \sqrt[12]{3^7}$

b) $\dfrac{\sqrt{5}}{\sqrt[3]{5}} = \dfrac{5^{\frac{1}{2}}}{5^{\frac{1}{3}}} = 5^{\frac{1}{2}-\frac{1}{3}} = 5^{\frac{3}{6}-\frac{2}{6}} = 5^{\frac{1}{6}} = \sqrt[6]{5}$

c) $\sqrt[3]{\sqrt[5]{10}} = \sqrt[15]{10}$

The next group of root rules deals with multiplication when the radicands are different. The root of a product is the product of the roots: $\sqrt[n]{a \cdot b} = \sqrt[n]{a} \cdot \sqrt[n]{b}$. The root of a quotient is the quotient of the roots: $\sqrt[n]{\dfrac{a}{b}} = \dfrac{\sqrt[n]{a}}{\sqrt[n]{b}}$.

If the roots are different, then you need to put everything under one radical. This will involve finding the least common multiple of the roots: $\sqrt[n]{a} \cdot \sqrt[m]{b} = \sqrt[mn]{a^m b^n}$.

The Real Deal

The root rules are these:

$$\sqrt[n]{a} \cdot \sqrt[m]{a} = \sqrt[mn]{a^{m+n}}, \ \frac{\sqrt[n]{a}}{\sqrt[m]{a}} = \sqrt[mn]{a^{m-n}},$$

$$\sqrt[m]{\sqrt[n]{a}} = \sqrt[mn]{a}, \ \sqrt[n]{a \cdot b} = \sqrt[n]{a} \cdot \sqrt[n]{b}, \ \sqrt[n]{\frac{a}{b}} = \frac{\sqrt[n]{a}}{\sqrt[n]{b}},$$

and $\sqrt[n]{a} \cdot \sqrt[m]{b} = \sqrt[mn]{a^m b^n}$.

It may be helpful to work out an example of this last simplification process.

Example 4: Simplify $\sqrt{5} \cdot \sqrt[3]{4}$.

Solution: $\sqrt{5} \cdot \sqrt[3]{4} = \sqrt[2\cdot3]{5^3 4^2} = \sqrt[6]{125 \cdot 16} = \sqrt[6]{2000}$.

Before you leave your roots, it's worth playing around with $\sqrt[n]{a^n}$. Notice that $\sqrt{2} \cdot \sqrt{2} = 2$. This is because the number that you have to square in order to get 2 is $\sqrt{2}$. Similarly, $\sqrt{3} \cdot \sqrt{3} = 3$. In general, $\sqrt{a} \cdot \sqrt{a} = a$. Since $\sqrt{a} \cdot \sqrt{a} = \sqrt{a^2}$, you can see that if a is a positive number, then $\sqrt{a^2} = a$. To explore $\sqrt{a^2}$ completely, you must also explore what happens when a is negative. If you evaluate $\sqrt{(-2)^2}$, you see that $\sqrt{(-2)^2} = \sqrt{(-2)(-2)} = \sqrt{4} = 2$. So $\sqrt{(-2)^2} = 2$, which is the absolute value of -2. In general, if a is a real number, then $\sqrt{a^2} = |a|$.

Let's turn to cube roots next. Since $\sqrt[3]{2} \cdot \sqrt[3]{2} \cdot \sqrt[3]{2} = 2$ (the number that you have to cube in order to get 2 is $\sqrt[3]{2}$), and $\sqrt[3]{2} \cdot \sqrt[3]{2} \cdot \sqrt[3]{2} = \sqrt[3]{2^3}$, you can write $\sqrt[3]{2^3} = 2$. It looks like $\sqrt[3]{a^3} = a$, at least if a is positive. But what happens if a is negative? To evaluate $\sqrt[3]{(-2)^3}$, you need to keep in mind the order of operations. One way to evaluate $\sqrt[3]{(-2)^3}$ is to write the cube root as a fractional exponent and use our properties of exponents: $\sqrt[3]{(-2)^3} = \left((-2)^3\right)^{\frac{1}{3}} = (-2)^1 = -2$.

Notice that if you had used this method to evaluate $\sqrt{(-2)^2}$, you would have gotten the wrong answer! Remember our warning about roots and negative numbers! In order to write the roots as exponents, you have to make sure that your expression makes sense, and even roots and negative numbers don't mix. Keep in mind that whenever you use a generalization of a pattern that you observe, you must keep on your toes to be sure that you haven't violated any of the rules.

Forewarning

The rules for roots are intended to be used when working with positive numbers.

Let's get back to your exploration of $\sqrt[n]{a^n}$. Notice that if you let a be any real number (negative or positive), the results depend on whether the root is even or odd. This idea can be generalized:

$$\sqrt[n]{a^n} = \begin{cases} a & n \text{ odd} \\ |a| & n \text{ even} \end{cases}$$

Of course, if you only want to deal with positive radicands, then it doesn't matter whether n is even or odd. While I am a firm believer in looking at the positive side of things, there are negative numbers out there as well, and you need to know how to deal with them, too.

A New Order (of Operations)

Now that you understand the relationship between exponentials, roots, and multiplication, you are ready for an updated list of the order of operations.

In our expanded list, exponentiation and roots come first, then multiplication and division (read left to right), and bringing up the rear are addition and subtraction. Of course, you can always throw parentheses in to mix things up. As usual, instructions inside parentheses are followed first, and work inside to outside.

Rationalizing the Denominator

Somewhere along the road, it became forbidden for radicals to exist in the denominator of a fraction. Whenever you see one, you have to get the radical out of the denominator. Getting rid of radicals can be tricky, so let's start you out slowly.

Square Roots

I've outlined some steps to help you rid the denominators of square roots.

1. If there are radicals in the denominator, combine them into one radical expression $\sqrt{a}$.

2. Multiply the fraction by a complicated version of 1: $\dfrac{\sqrt{a}}{\sqrt{a}}$. This will leave a factor of a in the denominator and effectively move the radical into the numerator.

3. Simplify the radical in the numerator and reduce the fraction as necessary.

It may help to see an example or two, just to get your bearings.

Example 5: Rationalize the denominators of the ratios:

a) $\dfrac{2}{\sqrt{3}}$

b) $\dfrac{15}{\sqrt{6}}$

c) $\dfrac{8 \cdot \sqrt{14}}{\sqrt{30}}$

Solution:

a) The $\sqrt{3}$ in the denominator must go away. Following the steps outlined, multiply the ratio by $\frac{\sqrt{3}}{\sqrt{3}}$: $\frac{2}{\sqrt{3}} \cdot \frac{\sqrt{3}}{\sqrt{3}} = \frac{2 \cdot \sqrt{3}}{3}$. The radical has been moved from the denominator into the numerator, and peace has been restored.

b) First, multiply the ratio by $\frac{\sqrt{6}}{\sqrt{6}}$, factor the numerator and denominator, and then cancel what you can: $\frac{15}{\sqrt{6}} \cdot \frac{\sqrt{6}}{\sqrt{6}} = \frac{15\sqrt{6}}{6} = \frac{5 \cdot \cancel{3} \cdot \sqrt{6}}{\cancel{3} \cdot 2} = \frac{5 \cdot \sqrt{6}}{2}$.

c) The first thing to do is multiply by $\frac{\sqrt{30}}{\sqrt{30}}$:

$$\frac{8 \cdot \sqrt{14}}{\sqrt{30}} \cdot \frac{\sqrt{30}}{\sqrt{30}} = \frac{8\sqrt{2 \cdot 7 \cdot 2 \cdot 5 \cdot 3}}{30}$$

Instead of multiplying the two radicands, factor them both. Then it is easier to identify the perfect squares and deal with smaller numbers. Pull out the perfect squares and reduce the fraction:

$$\frac{8\sqrt{2 \cdot 7 \cdot 2 \cdot 5 \cdot 3}}{30} = \frac{\cancel{2} \cdot 2 \cdot 2 \cdot 2\sqrt{7 \cdot 5 \cdot 3}}{\cancel{2} \cdot 3 \cdot 5} = \frac{8\sqrt{105}}{15}$$

Cube Roots

To get rid of the square root in the denominator, I advised you to multiply by 1, disguised as $\frac{\sqrt{a}}{\sqrt{a}}$. The goal was to get rid of the $\sqrt{a}$ in the denominator, and since $\sqrt{a} \cdot \sqrt{a} = a$ (assuming a is positive), multiplying by 1 disguised this way worked. If the denominator has a $\sqrt[3]{a}$ term, then multiplying by $\sqrt[3]{a}$ won't help, since $\sqrt[3]{a} \cdot \sqrt[3]{a} = \sqrt[3]{a^2}$. In order to get rid of $\sqrt[3]{a}$ in the denominator, you have to multiply by $\sqrt[3]{a^2}$ because $\sqrt[3]{a} \cdot \sqrt[3]{a^2} = \sqrt[3]{a^3} = a$. After that, the steps remain the same.

1. If there are radicals in the denominator, combine them into one radical expression $\sqrt[3]{a}$.

2. Multiply the fraction by a complicated version of 1: $\dfrac{\sqrt[3]{a^2}}{\sqrt[3]{a^2}}$. This will leave a factor of a in the denominator and pull the radical into the numerator.

3. Simplify the radical in the numerator and reduce the fraction as necessary.

Example 6: Rationalize the denominators of the ratios:

a) $\dfrac{3}{\sqrt[3]{4}}$

b) $\dfrac{\sqrt{6}}{\sqrt[3]{3}}$

Solution:

a) First, multiply by 1, disguised as $\dfrac{\sqrt[3]{4^2}}{\sqrt[3]{4^2}} = \dfrac{\sqrt[3]{16}}{\sqrt[3]{16}}$:
$\dfrac{3}{\sqrt[3]{4}} \cdot \dfrac{\sqrt[3]{16}}{\sqrt[3]{16}} = \dfrac{3\sqrt[3]{16}}{4}$. Next, you need to see if you can simplify $\sqrt[3]{16}$ by looking at the prime factorization of 16: $\sqrt[3]{16} = \sqrt[3]{2 \cdot 2 \cdot 2 \cdot 2} = 2\sqrt[3]{2}$. So $\dfrac{3}{\sqrt[3]{4}} = \dfrac{3\sqrt[3]{16}}{4} = \dfrac{3 \cdot \cancel{2}\sqrt[3]{2}}{2 \cdot \cancel{2}} = \dfrac{3\sqrt[3]{2}}{2}$. No further simplification is possible, so you are done.

b) First, multiply by $\dfrac{\sqrt[3]{3^2}}{\sqrt[3]{3^2}} = \dfrac{\sqrt[3]{9}}{\sqrt[3]{9}}$: $\dfrac{\sqrt{6}}{\sqrt[3]{3}} \cdot \dfrac{\sqrt[3]{9}}{\sqrt[3]{9}} = \dfrac{\sqrt[2]{6}\sqrt[3]{9}}{3}$.

Next, multiply the two radicals in the numerator and simplify:

$$\frac{\sqrt[2]{6}\sqrt[3]{9}}{3} = \frac{\sqrt[6]{6^3 9^2}}{3}$$

$$= \frac{\sqrt[6]{6 \cdot 6 \cdot 6 \cdot 9 \cdot 9}}{3}$$

$$= \frac{\sqrt[6]{3 \cdot 3 \cdot 3 \cdot 3 \cdot 3 \cdot 3 \cdot 3 \cdot 2 \cdot 2 \cdot 2}}{3}$$

$$= \frac{\cancel{3} \cdot \sqrt[6]{3 \cdot 2 \cdot 2 \cdot 2}}{\cancel{3}}$$

$$= \sqrt[6]{24}$$

The General Process

A ratio is considered simplified if there are no radicals in the denominator. If you have a denominator infested with an nth root, you still have to multiply by a disguised 1. The disguises just keep getting better and better. In order to get rid of a radical of the form $\sqrt[n]{a}$ in the denominator, you have to multiply by 1 disguised as $\dfrac{\sqrt[n]{a^{n-1}}}{\sqrt[n]{a^{n-1}}}$. Then follow steps 2 and 3 as before:

1. If there are radicals in the denominator, combine them into one radical expression $\sqrt[n]{a}$.

2. Multiply the fraction by a complicated version of 1: $\dfrac{\sqrt[n]{a^{n-1}}}{\sqrt[n]{a^{n-1}}}$. This will leave a factor of a in the denominator and pull the radical into the numerator.

3. Simplify the radical in the numerator and reduce the fraction as necessary.

Don't be fooled into thinking that a simplified expression actually looks simpler than it looked before. "Simple" is in the eye of the beholder.

The Least You Need to Know

- To simplify square roots, factor the radicand and pull out the perfect squares.

- Roots are just fractional exponents, so use your rules for exponents when simplifying them.

- Roots of products and quotients:
  $$\sqrt[n]{a} \cdot \sqrt[m]{a} = \sqrt[mn]{a^{m+n}}, \ \frac{\sqrt[n]{a}}{\sqrt[m]{a}} = \sqrt[mn]{a^{m-n}}, \text{ and}$$
  $$\sqrt[m]{\sqrt[n]{a}} = \sqrt[mn]{a}.$$

- Products and quotients of roots:
  $$\sqrt[n]{a \cdot b} = \sqrt[n]{a} \cdot \sqrt[n]{b}, \ \sqrt[n]{\frac{a}{b}} = \frac{\sqrt[n]{a}}{\sqrt[n]{b}}, \text{ and}$$
  $$\sqrt[n]{a} \cdot \sqrt[m]{b} = \sqrt[mn]{a^m b^n}.$$

- Rationalizing the denominator requires you to get rid of the radical in the denominator. If there is a $\sqrt{a}$ in the denominator, multiplying the expression by $\frac{\sqrt{a}}{\sqrt{a}}$ should do the trick.

Chapter 4

Setting the Record Straight

In This Chapter

- Cardinality of a set
- Describing sets
- Intersections and unions of sets
- Venn Diagrams

People first learned to count because they had something to count. If you want to count how many CDs you own, you would point to the first one and say "one." Then you would point to the next one and say "two." You would keep pointing and saying a new number until you either ran out of CDs or ran out of numbers. Because of our pattern of generating numbers, chances are that you will run out of CDs before you run out of numbers. The last number you say corresponds to the number of CDs that you own. Numbers got their start because people needed to know how much they had compared to everybody else. In this chapter, you'll become sophisticated collectors, and learn how to keep track of your inventory.

Sets and Cardinality

Mathematicians are collectors, and they keep their collections in sets. A set is any collection of things. Sets can be described, like the set of natural numbers, or they can be given names, like A or N. The things you put in sets are called *elements*.

Added Information

Classical algebra (old math) dealt with specific objects: numbers, polynomials, etc. Modern algebra (new math) emphasizes working with sets. The sets can contain numbers (as with classical algebra), but the actual contents of the set are irrelevant; the *relationships* between the objects of the set are what are important.

The elements of a set do not have to be numbers. In order to denote a set, enclose the elements in braces. For example, the set containing the author's favorite colors can be written {blue, red, purple}. You can call this set "Colors" and write Colors = {blue, red, purple}. The set containing the natural numbers can be written $\{1, 2, 3 \ldots\}$. The symbol used to denote the set of natural numbers is N, and you can write $N = \{1, 2, 3 \ldots\}$. The "..." in the braces indicates that the list never ends.

Normally, a set contains one-of-a-kind items. In our set Colors, you may want to list "blue" twice, to emphasize that it is the author's absolute favorite. If you are going to list your items more than once, or if you have duplicate items, you need to warn anyone who might look at your set, because in that situation different rules apply. The order that you list the items in your set doesn't matter. You just have to make sure that you have accounted for them all.

In order to say that an element is in your set, use the symbol $\in$. For example, if you wanted to say that 3 was in the set of natural numbers N, you could write $3 \in N$, or $3 \in \{1, 2, 3 \ldots\}$.

If you've just decided to start a collection, the first thing you need is a place to put your elements. That would be your set. Think of a set as a bag, or a shelf; it's the place where you will put your stuff. You are allowed to talk about your collection even if you haven't collected anything yet. The empty set (or the null set) is the set without any elements. It is often denoted $\{\ \}$, or $\varnothing$.

Forewarning

Be careful with your notation. The empty set is denoted $\{\ \}$, or $\varnothing$. You can't use both symbols at the same time if you want to talk about the empty set. The set $\{\varnothing\}$ is not the empty set, but rather a set that contains the empty set, kind of like a bag within a bag.

The number of elements in a set is called the *cardinality* of the set. For example, the set $\{2, 4, 6\}$ has cardinality 3. Cardinality is symbolized by $| \ |$ (the same symbols used for absolute value!). Using the symbol for cardinality, you can write the statement "the cardinality of the set $\{2, 4, 6\}$ is 3" using much less effort or ink: $|\{2,4,6\}| = 3$. In general, the cardinality of a set A is written $|A|$. The empty set has cardinality 0; this can be written as $|\{ \ \}| = 0$.

Describing Sets

Sometimes you can describe your sets using words rather than listing out the particular elements. For example, if our set consisted of all real numbers strictly between 0 and 1 (by "strictly" I mean that the numbers 0 and 1 are not in our set), then I would have a hard time trying to list all of the elements in our set. It would be easier for us to describe our set rather than try and write down all of the elements in that set. You could write this set as:

$$\{a : a \text{ is real}, a > 0 \text{ and } a < 1\}$$

Notice that I began my set with the letter a. Arbitrary elements in sets are often labeled with a letter, like a or x. The criteria to get into our set are listed after the colon. In order to be in our set, you have to be real, you have to be greater than 0 (you can't equal 0) and you have to be less than 1 (you can't equal 1).

In general, if you give a word description of a set, you should always follow this format: First give a generic way to refer to the elements in the set, then use a colon (:) to divide the bracketed region (however, the symbol police will not arrest you if you use a different symbol to divide the regions, like a vertical line |), and then give a complete description of the characteristics that all of the elements in your set have in common.

Combining Sets

Sometimes collectors acquire other people's collections. In this case, two sets are combined into one, and you have the union of the two sets. If A and B are two sets, then their *union*, written $A \cup B$ is the set of all elements that are in either set (or in both). For example, if $A = \{1, 2, 3\}$, and $B = \{2, 4, 6\}$, then $A \cup B = \{1, 2, 3, 4, 6\}$. Notice that 2 is in both A and B, but it is only listed once.

Sometimes collectors look at other people's sets to see what their sets have in common. The *intersection* of two sets, A and B, is the set of all of the elements that are in both A and B. The intersection of A and B is written $A \cap B$. For example, $A = \{1, 2, 3\}$, and $B = \{2, 4, 6\}$, then $A \cap B = \{2\}$. If two sets have nothing in common, then their intersection is the empty set $\varnothing$.

The Real Deal _____

$A \cup B$ is the set of elements that are in either A or B (or both). $A \cap B$ is the set of elements that are in both A and B.

The cardinality of A, B, $A \cup B$, and $A \cap B$ are related by the equation:

$$|A \cup B| = |A| + |B| - |A \cap B|$$

In other words, the number of elements in $A \cup B$ is equal to the sum of the number of elements in A and the number of elements in B, but since the elements that are in both would be counted twice, you have to subtract the elements in $A \cap B$.

Xtracts _____

The empty set is to set theory what 0 is to the real numbers. If you take the union of the empty set and any set A, the result is the set A. This sounds a lot like the fact that the sum of 0 and any number a is a ($0 + a = a$). If you take the intersection of the empty set and any set A, the result is the empty set. This sounds a lot like the fact that the product of 0 and any number a is 0 ($0 \times a = 0$).

If you start with a complete collection (called the *universal* set, and usually denoted U), and focus on only a few elements, then you are dealing with a *subset* of the original set. For example, if you have the set $U = \{1, 2, 3, 4, 5, 6, 7, 8, 9\}$, and all you care about are the even numbers, then the set of even numbers $\{2, 4, 6, 8\}$ is called a subset of U. If A is a subset of U, you would write $A \subseteq U$. The set A is a subset of U if all of the elements in A are elements of U. There are different kinds of subsets: subsets and proper subsets. A set can be a subset of itself, in which case you write $U \subseteq U$. A *proper* subset of a set U is a subset of U that is not equal to U. In other words, at least one element of U must be missing from every one of its proper subsets, so a set cannot be a proper subset of itself. If A is a proper subset of U, write $A \subset U$.

If you have a subset of a set, you can also talk about its complement. If A is a subset of U, then its complement, denoted $\bar{A}$ or A^c, is the set of all of the things in U that are not in A. If $U = \{1, 2, 3, 4, 5, 6, 7, 8, 9\}$ and $A = \{2, 4, 6, 8\}$, then $\bar{A} = \{1, 3, 5, 7, 9\}$. Notice that $A \cup \bar{A} = U$, and $A \cup \bar{A} = \varnothing$.

The Real Deal

$A \subseteq U$ if every element in A is also in U. If $A \subseteq U$, then $\bar{A}$ or A^c is the set of all elements in U that are not in A.

A set can have a finite cardinality (meaning that it contains a finite number of elements) or an infinite cardinality (meaning that it has an infinite number of elements). You probably have an idea of what it means to be finite, and you probably know the difference between a finite and infinite. Admittedly, those are difficult ideas to describe precisely. One way to characterize a finite set is that the part is less than the whole. In other words, a set A is finite if every proper subset, S, of A has a smaller cardinality. An infinite set would be a set that does not have this property. In other words, with an infinite set, there exists a proper subset that has the same cardinality as the original set. This is a mind-boggling idea, and is one that you can and should explore more.

Added Information

The mathematician Georg Cantor was instrumental in exploring the characteristics of infinity ∞. His conclusions both amazed and disturbed not only himself but also much of the mathematical community at the time.

Comparing Sets

It's only natural to want to compare your set to other people's sets. There are several ways to do this. You can either compare the size of the sets or the contents of the sets. Each of these types of comparisons has its own name.

Two sets are said to be *equal* if they match up, element for element. Remember, it doesn't matter what order the elements are listed in. So the sets $\{1, 2, 3, 4\}$ and $\{4, 3, 2, 1\}$ are equal. Notice that each one of these sets is a subset of the other. In fact, that's how you prove that two sets are equal to each other.

Two sets are said to be *equivalent* if they have the same cardinality. For example, the sets $\{1, 2, 3, 4\}$ and $\{13, 12, 11, 10\}$ are equivalent, since they have the same number of elements. They are not equal, since the elements themselves are different.

The Real Deal _____

Sets A and B are *equivalent* if $|A| = |B|$.
Sets A and B are *equal* if $A \subseteq B$ and
$B \subseteq A$.

Venn Diagrams

A Venn diagram is a visual way to show the relationship between two or more sets. Each set is represented by a circle, and all of the circles intersect somewhere. Elements are placed into the appropriate regions of the intersecting circles. Elements that are in overlapping regions belong to both sets (and are in the intersection). For example, if $U = \{1, 2, 3, 4, 5, 6, 7, 8, 9\}$, $A = \{2, 4, 6, 8\}$, and $B = \{1, 2, 3, 4, 5, 6\}$, then the Venn diagram for these sets is shown in Figure 4.1. If you were to construct a Venn diagram of the real numbers, it would look something like Figure 4.2.

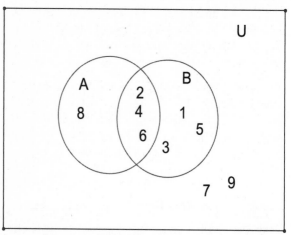

Figure 4.1

The Venn diagram for the sets $U = \{1, 2, 3, 4, 5, 6, 7, 8, 9\}$, $A = \{2, 4, 6, 8\}$, *and* $B = \{1, 2, 3, 4, 5, 6\}$.

Added Information

Venn diagrams were introduced in 1880 by John Venn. There is a stained-glass Venn diagram in Caius Hall at Cambridge University commemorating John Venn.

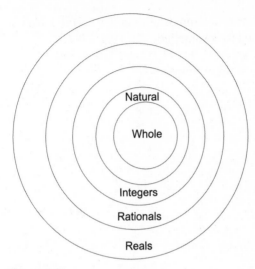

Figure 4.2
The Venn diagram for the real numbers.

The Least You Need to Know

- The cardinality of a set is the number of elements in that set.

- The union of a set is the set of all elements that are in either set; for example, A $\cup$ B is the set of elements that are in either A or B (or both).

- The intersection of a set is the set of all elements that are in both sets; for example, $A \cap B$ is the set of elements that are in both A and B.

- Sets A and B are equivalent if $|A| = |B|$; sets A and B are equal if $A \subseteq B$ and $B \subseteq A$.

- $A \subseteq U$ if every element in A is also in U (the universal set). If $A \subseteq U$, then its complement $\bar{A}$ or A^c is the set of all elements in U that are not in A.

- A Venn diagram is a visual way to show the relationship between two or more sets.

Chapter 5

Monomials and Polynomials

In This Chapter

- Manipulating monomials
- Using monomials to build polynomials
- Dividing polynomials by monomials
- Two factoring techniques: common factors and grouping

Now that you have reviewed numbers and sets, you are ready to start playing with algebraic expressions like monomials and polynomials. Numbers are the basis for what you can do in algebra. Realize that any symbol you see is just a generic way to write a number.

Monomials

A monomial is an algebraic expression that consists of one term. The term can have many parts, but all parts are multiplied (or divided) together. Monomials do not involve addition or subtraction. Examples of monomials include x, $3x$, $5a^2$, and $4mp^3$. The letters involved in a monomial are its variables, and the variables can be raised to any power.

The number in front of the variable is the *coefficient*. Whenever a coefficient is missing, you need to remember that the multiplicative identity is always lurking around in the shadows. You don't have to keep looking over your shoulder, since 1 will show itself whenever you ask. Whenever the power of a variable in a monomial is missing, you should assume that the exponent is 1. The coefficients of the monomials x, $3x$, $5a^2$, and $4mp^3$ are 1, 3, 5, and 4, respectively.

The *degree* of a monomial is the sum of the exponents of all of the variables involved in the monomial. The monomial $4x^4$ has degree 4, and the monomial $10x^2y^5$ has degree 7. Remember that $a^0 = 1$ if $a \neq 0$. Using this idea, a number can be seen as a monomial. For instance, the number 3 can be written as $3x^0$. Because $3 = 3x^0$, the degree of 3 is 0 (the power that x is raised to is 0). You can think of x^0 as a disguised form of 1. Because of this, any number can be interpreted as a monomial with degree 0.

The Real Deal

The **degree** of a monomial is the sum of the exponents of all of the variables that appear in the monomial. The degree of a number is 0.

Adding and Subtracting Monomials

In order to add or subtract monomials, the monomials must involve the *same* variables to the *same* powers. Only their coefficients can be different. For example, $1x^2y$ and $2x^2y$ can be added or subtracted since the variables involved (and their respective powers) in each monomial are the same; however, $1xy^2$ and $2xy$ cannot be combined since the powers of y are not the same. When you add or subtract monomials, the variables and their powers must be the same; only the coefficient changes.

Example 1: Combine the following monomials:

a) $13xyz + 15xyz$

b) $4x^2 - x^2$

c) $5x - 17x$

d) $4x - 3xy$

Solution:

a) $13xyz + 15xyz = 28xyz$

b) $4x^2 - x^2 = 3x^2$ (an unseen coefficient is really 1)

c) $5x - 17x = -12x$

d) $4x - 3xy = 4x - 3xy$ (the variables in each monomial are not the same, so there is nothing you can do!)

Notice that when you add or subtract monomials, your answer may or may not be a monomial, as seen in Example 1d (it is a polynomial, which will be explained in a later section).

Multiplying and Dividing Monomials

The reason I spent so much time talking about the rules of exponents in Chapter 2 is that those same rules apply in algebra. To multiply and divide monomials, treat each variable as a distinct number, and add or subtract the appropriate powers. The coefficients of each monomial are just multiplied or divided, depending on the instructions.

When multiplying or dividing monomials, it doesn't matter if the variables or their powers are the same (unlike adding and subtracting monomials). Since all of the variables within a monomial are multiplied together, dividing one monomial by another monomial will involve lots of cancellation. When you multiply or divide monomials, your answer will also be a monomial.

Example 2: Multiply the following monomials:

a) $(x^2)(x^5)$

b) $(3x^3)(5x^2)$

c) $(x^3y^2)(x^5y)$

d) $(-5xyz)(4xz^2)$

e) $(-4rs^4)(-5r^3s^2)(2r^4)$

Solution:

a) Just add the exponents:

$$(x^2)(x^5) = x^7$$

b) Multiply the coefficients and add the powers of x:

$$(3x^3)(5x^2) = 15x^5$$

c) Add the powers of the x's together and add the powers of the y's together:

$$(x^3y^2)(x^5y) = x^8y^3$$

d) Even though there is no y variable visible in the second monomial, you can pretend that y^0 is part of the second monomial; 1 can be introduced anywhere in a product since it is the multiplicative identity, and y^0 is just a complicated way of writing 1:

$$(-5xyz)(4xz^2) = (-5xyz)(4xy^0z^2) = -20x^2yz^3$$

e) Multiply each of the variables:

$$(-4rs^4)(-5r^3s^2)(2r^4) = 40r^8s^6$$

That was fun! Now let's practice our division skills.

Example 3: Divide the following monomials:

a) $\dfrac{x^5}{x^2}$

b) $\dfrac{30x^3}{5x^2}$

c) $\dfrac{x^3y^2}{x^5y}$

d) $\dfrac{(-5xyz)(4xz^2)}{2y^2z}$

e) $\dfrac{(-4rs^4)(r^3s^2)}{2r^4}$

Solution:

a) Subtract the exponents:

$$\frac{x^5}{x^2} = x^3$$

b) Divide the coefficients and subtract the exponents:

$$\frac{30x^3}{5x^2} = 6x$$

c) Subtract the exponents term by term:

$$\frac{x^3 y^2}{x^5 y} = x^{-2} y = \frac{y}{x^2}$$

Either form (with negative or positive exponents) is acceptable.

d) Simplify the numerator and then divide:

$$\frac{(-5xyz)(4xz^2)}{2y^2 z} = \frac{-20x^2 yz^3}{2y^2 z} = -10x^2 y^{-1} z^2 = \frac{-10x^2 z^2}{y}$$

Again, either form (with positive or negative exponents) is acceptable.

e) Simplify the numerator and then divide:

$$\frac{(-4rs^4)(r^3 s^2)}{2r^4} = \frac{-4r^4 s^6}{2r^4} = -2s^6$$

Exponentiation and Monomials

When a monomial is raised to a power, each variable in the monomial (including the coefficient) must be raised to that power. Because exponentiation is shorthand notation for multiplication, when you exponentiate a monomial, your answer will be a monomial. When you raise a power to a power, you multiply the two powers.

Example 4: Simplify:

a) $(x^3)^4$

b) $(4x^2)^2$

c) $(x^3y^2)^5$

d) $(3x^3yz^2)^3$

Solution:

a) $(x^3)^4 = x^{12}$

b) $(4x^2)^2 = 4^2x^4 = 16x^4$

c) $(x^3y^2)^5 = x^{15}y^{10}$

d) $(3x^3yz^2)^3 = 3^3x^9y^3z^6 = 27x^9y^3z^6$

Polynomials

The building blocks of polynomials are monomials. A polynomial is an expression that involves the sum of two or more monomials. You can add, subtract, multiply, and divide polynomials.

Since polynomials are made up of monomials, the notion of degree of a monomial can be extended to polynomials. The *degree* of a polynomial is the degree of the largest monomial that is contained within the polynomial. The polynomial $4x^3 - 2x + 1$ has degree 3. The polynomial $4xy^4 + 10x^4 + x^2y^4$ has degree 6 because the term x^2y^4 has degree 6.

The *leading coefficient* of a polynomial is the coefficient in front of the monomial with the highest degree. The leading coefficient of the polynomial $4x^3 - 2x + 1$ is 4.

Polynomials with degree 1, 2, and 3 have nicknames. Polynomials that have degree 1 are called linear polynomials. Polynomials with degree 2 are called quadratic polynomials. And polynomials with degree 3 are called cubic polynomials. There are nicknames for polynomials having other degrees, but that will have to wait until Algebra II!

Xtracts

Generic linear polynomials are of the form $ax + b$, generic quadratic polynomials are of the form $ax^2 + bx + c$, and generic cubic polynomials are of the form $ax^3 + bx^2 + cx + d$, where a, b, c, and d are real numbers.

Adding and Subtracting Polynomials

To add or subtract polynomials, follow the rules for adding and subtracting monomials. Find the parts of the polynomial that involve the same variables to the same powers, and do the addition or subtraction. Again, you can't do anything if the variables and powers don't match.

Example 5: Add or subtract the following polynomials:

a) $(x^2 + 2x + 4) + (2x^2 - 5x + 8)$

b) $(4x - 3y) + (5x + 8y)$

c) $(3r^3s + 2rs + 5rs^3) - (2r^3s + 4r^2s + 6rs^3)$

Solution:

a) Just add the like terms:

$$(x^2 + 2x + 4) + (2x^2 - 5x + 8) = 3x^2 - 3x + 12$$

b) Combine the like terms:

$$(4x - 3y) + (5x + 8y) = 9x + 5y$$

c) Distribute the negative sign, and then combine the like terms:

$$(3r^3s + 2rs + 5rs^3) - (2r^3s + 4r^2s + 6rs^3)$$
$$= (3r^3s + 2rs + 5rs^3) - 2r^3s - 4r^2s - 6rs^3$$

$$= r^3s + 2rs - 4r^2s - rs^3$$

Multiplying Polynomials

You can set up multiplication of polynomials in the same way that you set up multiplication of 2 or 3 digit numbers. There is a slight difference, however, that makes multiplication of polynomials a little easier. Recall that when you multiply two numbers, you have to keep track of the 1's place, the 10's place, etc. When you multiply polynomials, you don't have to keep track of the places. You just have to make sure that you multiply each term in one polynomial by each term in the other polynomial. When you multiply 24 and 12, you multiply the 4 and the 2 together, the 4 and the 1 together, the 2 and the 2 together, and the 2 and the 1 together:

$$
\begin{array}{r}
24 \\
\times \quad 12 \\
\hline
48 \\
+ \quad 240 \\
\hline
288
\end{array}
$$

Since $24 \cdot 12 = 288$, 12 and 24 are factors of 288. You will use the same terminology with polynomials.

When you multiply $x + 3$ and $2x + 7$ together, you will have to multiply each term in $x + 3$ by each term in $2x + 7$. There are several ways to visualize this, and I will show you two. One of them makes use of FOIL, as you saw in Chapter 1. The other is similar to the method of multiplying numbers that I just showed you. You should pick whichever style makes the most sense to you, and use it consistently.

Example 6: Multiply the polynomials:

a) $(2x + 4)(5x + 2)$

b) $(3x + y)(4x - 3y)$

c) $(4x + 2y + 3)(x + 4)$

Solution: Solutions are obtained using both methods:

a)
$$
\begin{array}{r}
2x + 4 \\
\times \quad 5x + 2 \\
\hline
4x + 8 \\
+ \quad 10x^2 + 20x \\
\hline
10x^2 + 24x + 8
\end{array}
$$

$(2x + 4)(5x + 2) = 10x^2 + 4x + 20x + 8 = 10x^2 + 24x + 8$

b)
$$
\begin{array}{r}
3x + y \\
\times \quad 4x - 3y \\
\hline
-9xy - 3y^2 \\
+ \quad 12x^2 + 4xy \\
\hline
12x^2 - 5xy - 3y^2
\end{array}
$$

$$(3x + y)(4x - 3y) = 12x^2 - 9xy + 4xy - 3y^2 =$$
$$12x^2 - 5xy - 3y^2$$

c)
$$
\begin{array}{r}
4x + 2y + 3 \\
\times \quad\quad x + 4 \\
\hline
16x + 8y + 12 \\
+ \quad 4x^2 + 2xy + 3x \\
\hline
4x^2 + 2xy + 19x + 8y + 12
\end{array}
$$

$$(4x + 2y + 3)(x + 4) = 4x^2 + 16x + 2xy + 8y + 3x + 12$$
$$= 4x^2 + 2xy + 19x + 8y + 12$$

In Example 6a you saw that $(2x + 4)(5x + 2) = 10x^2 + 24x + 8$. The terms $(2x + 4)$ and $(5x + 2)$ are called the *linear factors* of $10x^2 + 24x + 8$. Notice that the coefficient for the x^2 term came from the product of the leading coefficients (2 and 5) of each of the linear factors $((2x + 4)$ and $(5x + 2))$. Also, notice that the constant term (8) came from the product of the constant terms in each of the linear factors. That observation will come in handy when you learn how to factor polynomials.

Dividing Polynomials by Monomials

When a polynomial is divided by a monomial, you need to split the numerator (the polynomial) into the sum (or difference) of monomials. You will then have a string of monomials that are divided by monomials, and you have already discovered how to handle that situation. Perhaps an example or two will help illustrate the process.

Example 7: Divide the following polynomials by the specified monomial:

a) $(8x^3 + 3x) \div 4x$

b) $(16x^4 - 12x^3 + 20x^2) \div 4x^2$

Solution:

a) $\left(8x^3 + 3x\right) \div 4x = \dfrac{\left(8x^3 + 3x\right)}{4x} = \dfrac{8x^3}{4x} + \dfrac{3x}{4x} = 2x^2 + \dfrac{3}{4}$

b) $\left(16x^4 - 12x^3 + 20x^2\right) \div 4x^2 = \dfrac{\left(16x^4 - 12x^3 + 20x^2\right)}{4x^2}$

$$= \dfrac{16x^4}{4x^2} - \dfrac{12x^3}{4x^2} + \dfrac{20x^2}{4x^2}$$

$$= 4x^2 - 3x + 5$$

Factoring

Since $x^3(3x + 5) = 3x^4 + 5x^3$, you know that both x^3 and $(3x + 5)$ are factors of $3x^4 + 5x^3$. Now that you know how to multiply polynomials, it's time to go backward. Can you look at the polynomial $3x^4 + 5x^3$

and find its factors? The process of finding the factors of a polynomial is called *factoring*. I'll touch on some basic techniques now and go into more detail in Chapter 11.

Added Information

In 1923, Charles Babbage received a grant to build a prototype of a large "difference engine"—a machine that could factor polynomials. By 1934 he had designed what was called an "analytical engine," which was a forerunner of the modern computer. So factoring polynomials inspired the development of computers!

Common Factors

The first thing to do when trying to factor a polynomial is to look at each of the terms of the polynomial. Look for any monomials that appear in all of the terms of the polynomial. If there is a monomial that appears as a factor in each of the terms of the polynomial, divide the polynomial by that monomial. The resulting polynomial and the monomial are factors of the original polynomial. Look at the new polynomial, and see if there are any monomials that appear in all of the terms of this new polynomial. If there are, divide and conquer again. Keep going until there are no common monomial factors.

Example 8: Factor the following polynomials:

a) $7x^2 - 5x$

b) $4x^4 + 2x^3 + 6x^2$

c) $x^7 + x^5$

Solution:

a) x is a common factor in each term of the polynomial. Divide $7x^2 - 5x$ by x:

$$\frac{7x^2 - 5x}{x} = \frac{7x^2}{x} - \frac{5x}{x} = 7x - 5$$

Write your answer as the product of the monomial x and the new polynomial $7x - 5$: $7x^2 - 5x = x(7x - 5)$. If you distribute the x, the result will be what you started with.

b) $2x^2$ is a common factor in each term of the polynomial (don't just look at the variables, but pay attention to the coefficients as well!). Divide $4x^4 + 2x^3 + 6x^2$ by $2x^2$:

$$\frac{4x^4 + 2x^3 + 6x^2}{2x^2} = \frac{4x^4}{2x^2} + \frac{2x^3}{2x^2} + \frac{6x^2}{2x^2} = 2x^2 + x + 3$$

Write your answer as the product of the monomial $2x^2$ and the new polynomial $2x^2 + x + 3$: $4x^4 + 2x^3 + 6x^2 = 2x^2(2x^2 + x + 3)$. If you distribute the $2x^2$ term, the result will be what you started with.

c) x^5 is a common factor in each term of the polynomial. Divide $x^7 + x^5$ by x^5:

$$\frac{x^7 + x^5}{x^5} = \frac{x^7}{x^5} + \frac{x^5}{x^5} = x^2 + 1$$

Write your answer as the product of the monomial x^5 and the new polynomial $x^2 + 1$: $x^7 + x^5 = x^2(x^2 + 1)$. If you distribute the x^2 term, the result will be what you started with.

Grouping

Factoring by grouping can be useful, but don't rely on this method exclusively. One indication to try this method is if there are four terms involved in the polynomial.

Example 9: Factor the following polynomials:

a) $5x + xy - 2y - 10$

b) $x^2 - 5x + 3xy - 15y$

Solution:

a) Group the first two terms together and the last two terms together and pull out any common factors:

$$5x + xy - 2y - 10 = (5x + xy) - (2y + 10)$$
$$= x(5 + y) - 2(y + 5)$$
$$= (y + 5)(x - 2)$$

b) Group the first two terms together and the last two terms together and pull out any common factors:

$$x^2 - 5x + 3xy - 15y = (x^2 - 5x) + (3xy - 15y)$$
$$= x(x - 5) + 3y(x - 5)$$
$$= (x - 5)(x + 3y)$$

Sometimes you have to rearrange the terms and then try to factor by grouping. The key is to think about the terms involved and look for common factors between pairs of terms.

Forewarning

Factoring by grouping is not the most effective factoring strategy and should be used as a last resort; it should not be the first technique you try!

The Least You Need to Know

- To add or subtract monomials, they must have the same variables to the same powers; only their coefficients can be different.
- Use power rules when multiplying and dividing monomials.
- Use the FOIL method when multiplying polynomials.
- Dividing a polynomial by a monomial involves breaking the fraction up and then using power rules to simplify.
- The two factoring techniques for polynomials are pulling out a common factor and grouping.

Chapter 6

Rational Expressions

In This Chapter

- Long division with polynomials
- More simplification skills
- Multiplying and dividing rational expressions
- Adding and subtracting rational expressions

Rational expressions are fractions that involve polynomials. Because fractions imply division, rational expressions also involve division. In Chapter 5 you saw how to divide a polynomial by a monomial. You were able to use that skill when factoring out a common term in a polynomial. That was kind of like dividing a two- or three-digit number by a one-digit number. Now I'd like to show you how to divide one polynomial by another polynomial. To continue with the arithmetic analogy, it's kind of like dividing a three-digit number by a two-digit number using long division. You'll even use your familiar long division notation!

Added Information

Before the method of long division was developed, division was done using a method referred to as "galley" or "scratch." If you think long division is rough, spend some time exploring this older method ... you will be grateful that some anonymous mathematician took the time to develop long division.

Dividing Polynomials by Polynomials

Dividing a polynomial by another polynomial is similar to long division with natural numbers. Long division is a process developed to divide two numbers and obtain the answer one digit at a time. Since you are working with polynomials rather than natural numbers, you must make use of the degree of the polynomials involved in the division. The polynomial in the numerator is the *dividend*, and the polynomial in the denominator is the *divisor*. A systematic approach to division is as follows:

1. Arrange the terms (the monomials) of both polynomials in descending order according to their degrees. If the polynomial has gaps between consecutive terms, like the polynomial $3x^3 + 2x - 1$ (it is missing an x^2 term) leave a space.

2. Divide the first term (or the leading term) of the divisor into the first term (or the leading term) of the dividend. Place this answer above the long division symbol.

3. Multiply the divisor by the expression just written above the division symbol and place this answer directly below the dividend. Line up like terms. The polynomial written here is called the *quotient*.

4. Subtract this new polynomial from the dividend.

5. If the degree of the result from the subtraction is less than the degree of the divisor, you are finished. If it is not, repeat steps 2 through 4 until you are left with a polynomial having a smaller degree than the divisor. The leftover polynomial is called the *remainder*.

The Real Deal

When you divide one polynomial by another, the polynomial in the numerator is the **dividend,** and the polynomial in the denominator is the **divisor**. The polynomial written above the division symbol is the **quotient,** and the polynomial left over when you are done dividing is the **remainder**.

If the remainder is zero, then the division is exact, and the quotient is the polynomial written across the top of the long division symbol. If the remainder is not zero, then the quotient is the polynomial written across the top of the long division symbol plus the remainder divided by the divisor.

Example 1: Find the following quotients:

a) $(4x^2 + 10x + 4) \div (x + 2)$

b) $(6x^2 + 12x + 1) \div (2x + 3)$

c) $(x^3 + 1) \div (x + 1)$

Solution: The steps involved are these: divide, multiply, subtract, divide, multiply, subtract, divide, multiply, subtract …

a) First, write the problem so that you are ready for the long division:

$$x + 2 \overline{)4x^2 + 10x + 4}$$

Then, follow the steps until the degree of the remainder is less than the degree of the divisor; don't stop until you have a remainder of degree 0 … which means it must be a number. The leading term of the dividend is $4x^2$ and the leading term of the divisor is x. Since $\dfrac{4x^2}{x} = 4x$, $4x$ goes on top of the long division symbol:

$$x + 2 \overline{)\begin{array}{l} 4x \\ 4x^2 + 10x + 4 \end{array}}$$

Next, multiply the divisor by $4x$ and put that underneath the dividend:

$$4x(x + 2) = 4x^2 + 8x$$

$$\begin{array}{r} 4x \\ x+2\overline{\smash{\big)}4x^2 + 10x + 4} \\ 4x^2 + 8x \end{array}$$

Now subtract the two polynomials and bring down the next term:

$$\begin{array}{r} 4x \\ x+2\overline{\smash{\big)}4x^2 + 10x + 4} \\ (-)\underline{4x^2 + 8x } \\ 2x + 4 \end{array}$$

Forewarning

Remember to distribute the negative sign when you are subtracting one polynomial from the other!

The remainder at this stage is $2x + 4$ … the degree of the remainder is 1 and the degree of the divisor $(x + 2)$ is also 1, so you can't stop yet. Follow the same steps as before, this time focusing on $2x + 4$. The leading term of the dividend is $2x$ and the leading term of the divisor is x. Divide $2x$ by x to get 2, the next term to write above the long division symbol. Multiply the divisor $(x + 2)$ by 2 and write that polynomial below the previous remainder. Then subtract to get the new and improved remainder:

$$
\begin{array}{r}
4x + 2 \\
x + 2 \overline{)4x^2 + 10x + 4} \\
(-)\underline{4x^2 + 8x} \\
2x + 4 \\
(-)\underline{2x + 4} \\
0
\end{array}
$$

Since the remainder is 0, you know that $(x + 2)$ divides evenly into $4x^2 + 10x + 4$, and $\frac{4x^2 + 10x + 4}{x + 2} = 4x + 2$. The polynomial $4x + 2$ is called the *quotient*.

b) To perform the division $(6x^2 + 12x + 1) \div (2x + 3)$, follow the same steps as in the last example:

$$\frac{6x^2}{2x} = 3x,$$

$$
\begin{array}{r}
3x \\
2x + 3 \overline{)6x^2 + 12x + 1} \\
(-)\underline{6x^2 + 9x} \\
3x + 1
\end{array}
$$

At this point, the remainder is $(3x + 1)$, and the divisor is $2x + 3$. Both of these polynomials have degree 1, so you have to keep going:

$$\frac{3x}{2x} = \frac{3}{2},$$

$$\begin{array}{r} 3x + \frac{3}{2} \\ 2x+3\overline{\smash{)}6x^2 + 12x + 1} \\ (-)\underline{6x^2 + \ 9x} \\ 3x + 1 \\ (-)\underline{3x + \frac{9}{2}} \\ -\frac{7}{2} \end{array}$$

$$\begin{array}{r} 3x + \frac{3}{2} \\ 2x+3\overline{\smash{)}6x^2 + 12x + 1} \\ (-)\underline{6x^2 + \ 9x} \\ 3x + 1 \\ (-)\underline{3x + \frac{9}{2}} \\ -\frac{7}{2} \end{array}$$

The degree of the remainder is 0 (the remainder is not zero, but its degree is!), which is less than the degree of the divisor, so you are done. So:

$$\frac{6x^2 + 12x + 1}{2x + 3} = 3x + \frac{3}{2} - \frac{\frac{7}{2}}{2x+3} = 3x + \frac{3}{2} - \frac{7}{4x+6}$$

The quotient is $3x + \frac{3}{2}$ and the remainder is $-\frac{7}{2}$.

c) Now that you know the process, I'll just show you how you would solve this problem. Notice that the polynomial $x^3 + 1$ is missing the x^2 and x terms, so you'll need to fill those in when you solve the problem. Don't be afraid to close the book now (bookmark this page, though) and work this problem out on your own before looking at our solution:

$$\begin{array}{r} x^2 - x + 1 \\ x+1\overline{)x^3 + 0x^2 + 0x + 1} \\ (-)\underline{x^3 +\ x^2} \\ -x^2 +\ 1 \\ (-)\underline{-x^2 -\ x} \\ x+1 \\ (-)\underline{x+1} \\ 0 \end{array}$$

The quotient is $x^2 - x + 1$ and the remainder is 0.

Rational Expressions

At this point, I hope you are getting comfortable with polynomials. Most of the rules developed for manipulating polynomials come from our rules for manipulating real numbers. Now it's time to generalize the rules for fractions. A fraction is a ratio of two numbers, and a *rational expression* is our fancy name for a ratio of two polynomials. The ratios $\dfrac{x+3}{x-2}$ and $\dfrac{x^2 + 2x + 4}{x-1}$ are examples of rational expressions. There is one restriction regarding fractions: you cannot divide by zero. That same rule applies to rational expressions. You are not allowed to have a zero in the denominator. So the rational expression $\frac{x+3}{x-2}$ only makes sense if $x \neq 2$, and the rational expression $\frac{x^2 + 2x + 4}{x - 1}$ only makes sense if $x \neq 1$. You must always keep these restrictions in the back of your head. For the purpose of this book, assume that the rational expressions are valid only for values of the variable for which the denominator is nonzero.

A fraction like $\frac{15}{12}$ begs to be simplified. Not only is the fraction greater than 1, but it is not in reduced form. To put this fraction into reduced form, you would factor the numerator and denominator and look for common factors to cancel. The rational expressions $\frac{x+3}{x-2}$ and $\frac{x^2 + 2x + 4}{x - 1}$ are similar to $\frac{15}{12}$; there is an opportunity to do the division and write these rational expressions in a more formal way, and there is also an opportunity to look for common factors to cancel.

Simplifying Rational Expressions

You won't be doing any long division here; the goal in this section is to cancel common factors and put the rational expressions in reduced form. That means that you need to factor the numerator and factor the denominator.

Example 2: Simplify the rational expressions:

a) $\frac{x^2 + 8x + 15}{x^2 - x - 12}$

b) $\frac{x^2 - 3x - 4}{x^2 + x - 20}$

Solution: It's great to have another chance to practice factoring!

a) $\dfrac{x^2 + 8x + 15}{x^2 - x - 12} = \dfrac{\cancel{(x+3)}\,(x+5)}{\cancel{(x+3)}\,(x-4)} = \dfrac{(x+5)}{(x-4)}$

b) $\dfrac{x^2 - 3x - 4}{x^2 + x - 20} = \dfrac{\cancel{(x-4)}\,(x+1)}{\cancel{(x-4)}\,(x+5)} = \dfrac{(x+1)}{(x+5)}$

Multiplying Rational Expressions

To multiply rational expressions, follow the same rules that you had for multiplying fractions. You multiply the numerators of the rational expressions together and the denominators of the rational expressions together. Try to cancel as much as possible. If the rational expressions are not factored, you will need to factor them.

Example 3: Multiply the following rational expressions:

a) $\dfrac{x+1}{x^2 + x - 2} \cdot \dfrac{x-1}{x^2 + x}$

b) $\dfrac{x-3}{x^2 + 5x + 6} \cdot \dfrac{x+2}{x^2 - 9}$

Solution: Factor all of the terms, try to cancel as much as you can, and then create one big rational expression.

a) $\dfrac{x+1}{x^2 + x - 2} \cdot \dfrac{x-1}{x^2 + x} = \dfrac{\cancel{(x+1)}}{(x+2)\,\cancel{(x-1)}} \cdot \dfrac{\cancel{(x-1)}}{x\,\cancel{(x+1)}} =$

$\dfrac{1}{x(x+2)}$

b) $\dfrac{x-3}{x^2+5x+6} \cdot \dfrac{x+2}{x^2-9} = \dfrac{\cancel{(x-3)}}{(x+3)\cancel{(x+2)}} \cdot \dfrac{\cancel{(x+2)}}{\cancel{(x-3)}(x+3)} =$

$\dfrac{1}{(x+3)^2}$

Dividing Rational Expressions

To divide rational expressions, just follow the same process for dividing fractions: invert and multiply. Of course, a little factoring along the way lets you cancel and simplify.

Xtracts

When dividing rational expressions, invert the denominator and multiply.

Example 4: Divide the following rational expressions:

a) $\dfrac{x^2+x-2}{x+3} \div \dfrac{x+2}{x-1}$

b) $\dfrac{x^2+4x+3}{x-1} \div \dfrac{x^2-1}{x+2}$

Solution: The key is to invert and multiply (don't forget to factor!):

a) $\dfrac{x^2 + x - 2}{x + 3} \div \dfrac{x + 2}{x - 1} = \dfrac{x^2 + x - 2}{x + 3} \cdot \dfrac{x - 1}{x + 2}$

$$= \dfrac{(x+2)(x-1)}{x+3} \cdot \dfrac{(x-1)}{(x+2)}$$

$$= \dfrac{(x-1)^2}{x+3}$$

b) $\dfrac{x^2 + 4x + 3}{x - 1} \div \dfrac{x^2 - 1}{x + 2} = \dfrac{x^2 + 4x + 3}{x - 1} \cdot \dfrac{x + 2}{x^2 - 1}$

$$= \dfrac{(x+3)(x+1)}{x-1} \cdot \dfrac{x+2}{(x-1)(x+1)}$$

$$= \dfrac{(x+3)(x+2)}{(x-1)^2}$$

Adding and Subtracting Rational Expressions

You could probably write the rules for adding and subtracting rational expressions. All you have to do is think about how you add and subtract fractions. Get a common denominator and then add the numerators. You may have to FOIL the numerators before you can add them.

Xtracts

In order to add rational expressions, both expressions must have the same denominator. Then you just add the numerators and simplify.

Example 5: Combine the following rational expressions:

a) $\dfrac{3}{x+2} + \dfrac{1}{x-2}$

b) $\dfrac{1}{x+2} - \dfrac{4}{x-3}$

c) $\dfrac{x+2}{x-5} + \dfrac{x+1}{x-2}$

Solution:

a) Get a common denominator by multiplying each expression by a disguised 1:

$$\frac{3}{x+2} + \frac{1}{x-2} = \frac{3}{x+2} \cdot \frac{(x-2)}{(x-2)} + \frac{1}{x-2} \cdot \frac{(x+2)}{(x+2)}$$

Distribute the constants in the numerator, but keep the denominator factored:

$$\frac{3(x-2)}{(x+2)(x-2)} + \frac{1(x+2)}{(x-2)(x+2)} =$$

$$\frac{3x-6}{(x+2)(x-2)} + \frac{x+2}{(x-2)(x+2)}$$

Now that the denominators are the same, you can add the two rational expressions:

$$\frac{3x-6}{(x+2)(x-2)} + \frac{x+2}{(x-2)(x+2)} = \frac{4x-4}{(x-2)(x+2)}$$

The last step is to try to simplify the rational expression by factoring the numerator (if possible) and canceling:

$$\frac{4x-4}{(x-2)(x+2)} = \frac{4(x-1)}{(x-2)(x+2)}$$

b) Follow the same steps as in the example above:

$$\frac{1}{x+2} - \frac{4}{x-3} = \frac{1}{(x+2)} \cdot \frac{(x-3)}{(x-3)} - \frac{4}{(x-3)} \cdot \frac{(x+2)}{(x+2)}$$

$$= \frac{1(x-3)}{(x+2)(x-3)} - \frac{4(x+2)}{(x-3)(x+2)}$$

$$= \frac{x-3}{(x+2)(x-3)} - \frac{(4x+8)}{(x-3)(x+2)}$$

$$= \frac{x-3-4x-8}{(x+2)(x-3)}$$

$$= \frac{-3x-11}{(x+2)(x-3)}$$

c) Follow the same steps as in the examples above:

$$\frac{x+2}{x-5} + \frac{x+1}{x-2} = \frac{(x+2)}{(x-5)} \cdot \frac{(x-2)}{(x-2)} + \frac{(x+1)}{(x-2)} \cdot \frac{(x-5)}{(x-5)}$$

$$= \frac{(x+2)(x-2)}{(x-5)(x-2)} + \frac{(x+1)(x-5)}{(x-2)(x-5)}$$

$$= \frac{x^2-4}{(x-5)(x-2)} + \frac{x^2-4x-5}{(x-2)(x-5)}$$

$$= \frac{2x^2-4x-9}{(x-5)(x-2)}$$

The numerator and the denominator do not share any common factors, so there's no more cancellation that you can do.

The Least You Need to Know

- When you divide a polynomial by another polynomial, you must use long division.
- When you multiply rational expressions, multiply the numerators together and multiply the denominators together.
- When you divide rational expressions, invert the denominator and multiply.
- When you add or subtract rational expressions, you have to first find the common denominator.

Equations and Equalities

In This Chapter

- Solving equations
- Equivalence relations
- Playing algebraic games
- Absolute value

There are many ways to disguise numbers. For example, you can represent the number 2 as $\frac{4}{2}$, as $5 + 3 - 6$, or as $\frac{3+5}{4}$. If $x = 5$, you could also disguise it as $\frac{3+x}{4}$. All of these representations are equivalent since they all mean the same thing: 2. You can use an equation to show that these expressions are all equivalent. Actually, you can write several equations, but I'll just show you one: $\frac{3+x}{4} = 2$. Equations are used to show that two algebraic expressions are equivalent.

Once you have an equation that involves a variable, like the equation $\frac{3+x}{4} = 2$ for example, it is natural to want to know the value of the variable, in this case x. The value of the variable is a secret, and the goal is to be one of the people "in the know." Your mission will be to join the few, the proud, the people who know x.

Equations and Expressions

An equation is a mathematical statement that says two mathematical expressions are equal. In English, you start with an = symbol and put some stuff to the left of it and some stuff to the right of it. You need to make sure that you put an equivalent amount of stuff on each side. The stuff on each side probably won't look identical in form, but it *must* be identical in content. Whenever you write an equation, make sure that the expressions on each side represent the same thing.

You have seen equations earlier in this book. For instance, the statement $1 + 1 = 2$ is an equation that relates numbers to numbers. Equations can also relate a number to a symbol. The equation $x = 5$ can be used to simplify algebraic expressions, as you saw in Chapter 1. Equations can involve more than one variable, and can have numbers all over the place: $x - y + 4 = 7$ is an example of an equation with two variables (x and y) and two different numbers (4 and 7). You can make equations as simple or as complicated as you like.

Once you have an equation, you usually have to do something with it. And of course there are rules that you have to follow. You will get to work with equations soon enough, but before I tell you the rules of the game, you will need to solidify your knowledge of equality.

Equality in Relationships

Equality is a relation. A relation is just something that relates some things to other things. In algebra, you will relate numbers to symbols. Equality is not the only relation that you have encountered in this book. You have also seen the relations "less than" and "greater than." Relations can relate numbers to numbers, and can be used to relate numbers to symbols as well.

When used with real numbers, equality has some pretty nice properties. You have been using equality for many years and have taken many of these properties for granted. It's about time that you see equality for what it really is.

Equality is *reflexive*. That's a fancy way of saying that anything is equal to itself. It is what it is. You can write that mathematically as this: if a is a real number, then $a = a$.

Equality is *symmetric*. By that I mean that if $a = b$, then $b = a$. It doesn't matter how you look at it; frontward or backward you still have the same thing. For example, a dozen = 12, and 12 = a dozen (except in a bakery, but don't get us started on baker's math!).

Perhaps a better illustration would be this: if $1 + 1 = 2$, then $2 = 1 + 1$.

Equality is *transitive*. If $a = b$ and $b = c$, then $a = c$. This can be illustrated by the following example: if $3 + 5 = 8$ and $8 = 4 + 4$, then $3 + 5 = 4 + 4$.

The properties of equality are certainly reasonable, and not unfamiliar. Because equality possesses these three properties, it is called an *equivalence relation*. An equivalence relation is a relation that has the reflexive, symmetric, and transitive properties. Not all relations have those three properties.

The Real Deal

An **equivalence relation** is a relation that has the reflexive, symmetric, and transitive properties.

Example 1: Is the relation *greater than* an equivalence relation?

Solution: In order for a relation to be an equivalence relation, it must have the reflexive, symmetric, and transitive properties. Let's see which of these the relation *greater than* has:

> Reflexive: Is a number greater than itself? In other words, is $a > a$? Take a specific example: Is $4 > 4$? No. So *greater than* does not have the reflexive property.

Symmetric: If $a > b$, then is $b > a$? Take some specific numbers: If $8 > 5$, is $5 > 8$? It would be a strange world if it were. So *greater than* does not have the symmetric property.

Transitive: If $a > b$ and $b > c$, then is $a > c$? This one seems to work. If $8 > 5$ and $5 > 3$, is $8 > 3$? Yes. So *greater than* has the transitive property.

So *greater than* cannot be an equivalence relation, since it doesn't have the reflexive or symmetric properties. To its credit, it does at least have the transitive property.

Xtracts

Equality is an equivalence relation; greater than and less than are not.

Algebraic Games People Play

Equality represents balance. When playing around with equations, you must be careful not to upset the balance of power, or you'll have an all-out war. When manipulating equations, remember the golden rule: Do unto one side as you do unto the other. Often times, the goal is to isolate the variable. This is referred to as solving for the variable. The variable is isolated when the variable appears on only one side of the equation, its coefficient is 1, and only numbers or other variables are on the other side. When you've isolated your variable, and you have *an answer*, the last thing you should do is check it to see if it is correct. Checking your work usually involves the "plug and chug" method. Don't confuse it with the Fear Factor "plug and chug" method: plugging your nose while drinking something gross. The mathematical version requires you to "plug" the value that you think x is and then "chug" through the calculations. Evaluate each of the original expressions in the equation and see if you get the same result. If you do, you are right, and if you don't then you have a (different) problem ... find the error of your ways.

The Old Add-'Em/Subtract-'Em Trick

Anything that you do to an algebraic equation must maintain balance. You are allowed to add (or subtract) things to (or from) an algebraic equation, as long as you do the same thing to both sides. The keys to playing the algebra game are these: Be fair to both sides and remember that your objective is to isolate the variable.

Example 2: Solve for x: $x + 4 = 8$

Solution: You want x by itself, so you want to get rid of the 4 that is on the same side as the x. You want to add something to both sides of the equation so that the 4 goes away: The best thing to add to both sides would be -4: $x + 4 - 4 = 8 - 4$. Now do the subtraction: $x + 0 = 4$. So $x = 4$. You have *an answer*, so check it: Go to the left side of the equation and replace all of the x's with 4's: $4 + 4$. The right-hand side is 8. The two expressions are equal, so your answer works!

Example 3: Solve for x: $2x + 4 = x + 10$

Solution: You want x by itself. Notice that there are x's and numbers on each side of the equation. Bring the x's over to one side (the left) and move the 4 to the other side (the right). Take your time as you make your moves:

$$2x + 4 = x + 10$$
$$2x + 4 - x = x + 10 - x$$
$$x + 4 = 10$$
$$x + 4 - 4 = 10 - 4$$
$$x = 6$$

The last step is to check your work: The left-hand side of the original equation (with x's replaced with 6's) reads $2 \cdot 6 + 4$, which is 16. The right-hand side of the original equation reads $6 + 10$, which is also 16! The answer makes sense.

The Old Multiply-'Em/Divide-'Em Trick

Now that you've stretched your muscles, it's time to put some weight on the machine.

Example 4: Solve for x: $4x = 8$

Solution: The variable is isolated, but its coefficient is not 1; the coefficient is 4. How can you turn a $4x$ into a $1x$ (or just x)? You could subtract $3x$ from both sides, but then you won't have an isolated variable: x will be on both sides of the equation. You could multiply the $4x$ by $\frac{1}{4}$: $\frac{1}{4}(4x) = \left(\frac{1}{4} \cdot 4\right)x = 1x$. Thank goodness that multiplication is associative! Remember that if you do something to one side of the equality, you have to do it to the other side. So you should multiply both sides of the equation by $\frac{1}{4}$:

$$4x = 8$$

$$\left(\frac{1}{4}\right)4x = \left(\frac{1}{4}\right)8$$

$$1x = 2$$

So $x = 2$. Notice that multiplying by $\frac{1}{4}$ is the same thing as dividing by 4. The last thing to do is check your work: The left-hand side of the original equation reads $4 \cdot 2$, which is 8. The right-hand side is already 8, so your answer is correct!

The old multiply-'em trick and the old divide-'em trick are really the same trick going by two different names. A rose, by any other name, would smell as sweet ... the same holds true for these tricks.

Putting the Pieces Together

Sometimes problems require you to gather terms together (using the add-'em/subtract-'em trick) and then to make the variable coefficient equal to 1 (using the multiply-'em/divide-'em trick). If you run into those kinds of problems, apply the tricks in that order (gather terms together, and then make the coefficient 1) and you can't go wrong. Just don't forget to check your work.

Because I'm old (but wise) and set in my ways, I am in the habit of moving the x's to the left and the numbers to the right. It doesn't matter which side you move things to, as long as you separate numbers from the variables (the mathematical wheat from the chaff). Because I tend to prefer moving the x's to the left, there are times that I have to deal with negative numbers. I'm not afraid of a little negative sign or two, and you shouldn't be either. I recommend that you mix things up a little, avoid getting into algebraic ruts, and try to see the positive side of working with negative numbers!

Example 5: Solve for x: $3x - 8 = 5x - 2$

Solution: Move the x's to the left and the numbers to the right:

$$3x - 8 = 5x - 2$$
$$3x - 8 + 8 = 5x - 2 + 8$$
$$3x = 5x + 6$$
$$3x - 5x = 5x + 6 - 5x$$
$$-2x = 6$$

Now that the x's are on one side and the numbers are on the other, it's time to get the coefficient in front of x equal to 1. You need to multiply $-2x$ by something so that you end up with $1x$. In order to take care of the negative, you'll need to multiply by something negative. And to turn the 2 into a 1, you'll need to multiply by $\frac{1}{2}$. Putting it all together, you'll want to multiply both sides of the equation by $-\frac{1}{2}$:

$$-2x = 6$$
$$\left(-\frac{1}{2}\right)(-2x) = \left(-\frac{1}{2}\right)6$$
$$x = -3$$

The last step: Check your work. The left-hand side of the original equation reads like this: $3 \cdot (-3)-8$, or -17. The right-hand side of the equation reads like this: $5 \cdot (-3)-2$, which is also -17! You are on a roll!

 Forewarning

Remember your order of operations: multiply first, then subtract.

Example 6: Solve for x: $\frac{1}{3}x - 3 = \frac{1}{4}x + 2$

Solution: Don't let the fractions bother you. Follow the same procedure as before, and don't forget to breathe. First, isolate the variable:

$$\frac{1}{3}x - 3 = \frac{1}{4}x + 2$$

$$\frac{1}{3}x - 3 + 3 = \frac{1}{4}x + 2 + 3$$

$$\frac{1}{3}x = \frac{1}{4}x + 5$$

$$\frac{1}{3}x - \frac{1}{4}x = \frac{1}{4}x - \frac{1}{4}x + 5$$

$$\frac{1}{3}x - \frac{1}{4}x = 5$$

$$\frac{4}{12}x - \frac{3}{12}x = 5$$

$$\frac{1}{12}x = 5$$

Next, make the coefficient in front of the variable equal to 1:

$$\frac{1}{12}x = 5$$

$$(12)\frac{1}{12}x = 12 \cdot 5$$

$$1x = 60$$

So $x = 60$. You are finished when your work is checked. The left-hand side of the original equation reads $\frac{1}{3} \cdot 60 - 3$, or 17. The right-hand side of the original equation reads $\frac{1}{4} \cdot 60 + 2$, which is also 17! It works.

Linear Equations in One Variable

The problems you worked out earlier are examples of linear equations in one variable. A linear equation in one variable is an equation of the form $ax + b = c$ or $ax + b = cx + d$. Each side can involve the variable multiplied by a real number and added to a real number. Linear equations in one variable will always have either (a) no real solutions, (b) all real numbers as solutions, or (c) one real number as a solution.

The original equation has no solution if, after simplifying the equation, you end up with an equation that is false, like the equation $x = x + 7$. There is no real number that is unchanged when you add 7 to it.

All real numbers are solutions to the original equation if after simplifying the equation the result is an identity, like the equation $x = 4x - 3x$. All real numbers satisfy that equation.

If neither of these two situations occurs, then the original equation has one unique solution.

Determining which of these conditions holds is straightforward. All linear equations can be simplified into the form $ax = b$ for some real numbers a and b. If $a = 0$ and $b \neq 0$, then there is no solution. If $a = 0$ and $b = 0$, then all real numbers are solutions. If $a \neq 0$ and $b \neq 0$, then there is one unique solution: $x = \dfrac{b}{a}$.

Equations Without Numbers

Dealing with equations that only involve numbers (like $1 + 2 + 3 = 6$) can be comforting. Dealing with equations that involve only one variable and some numbers can be a little intimidating, but with a strategy for problem-solving they should pose no problem. But what about equations that only involve variables, and no numbers at all? Then you may be completely out of your comfort zone. Pull up a chair and get comfortable.

Variables are just substitutes for numbers. In fact, sometimes when you have too many variables, it may be easier for you to pretend that they are just numbers. Be careful not to think of them as either 0 or 1, since these two numbers can disappear and reappear at the drop of a hat. Think about how you work with numbers, and then work with the variables in the same way.

Example 7: Solve for r: $r \cdot t = d$

Solution: To solve for r, you need to isolate it from the other variables hanging around it. In order to isolate r from t, look at how they are tangled up, and then untangle them. Notice that r and t are multiplied together. The way to untangle them is to divide both sides of the equation by t, and then cancel:

$$r \cdot t = d$$

$$\frac{r \cdot \cancel{t}}{\cancel{t}} = \frac{d}{t}$$

$$r = \frac{d}{t}$$

The variables in this equation have specific interpretations: r stands for rate, t stands for time, and d stands for distance. This equation is read "rate times time equals distance" and can be used to answer the age-old question of "are we there yet?"

Example 8: Solve for P: $\frac{PV}{R} = nT$

Solution: Don't let all the letters confuse you. P is the important variable in this problem (since that's what you were told to solve). In order to isolate P, you need to move the V and the R over to the other side. Since P is multiplied by V, untangling P and V involves dividing by V. P is also divided by R, so untangling P and R is done by multiplying by R:

$$\frac{PV}{R} = nT$$

$$\frac{1}{V} \cdot \frac{PV}{R} = \frac{1}{V} \cdot nT$$

$$R \cdot \frac{P}{R} = R \cdot \frac{nT}{V}$$

$$P = \frac{nRT}{V}$$

Xtracts

$PV = nRT$ is a well-known formula in chemistry. It gives the relationship between the pressure exerted by an ideal gas, the volume that the gas occupies, and the temperature of the gas. If you are curious about this formula and what n and R represent, look up the Ideal Gas Law on the Internet.

Absolute Value

To solve an equation that involves an absolute value, the first thing you need to do is to isolate the absolute value. Remember that the absolute value of a number is always positive. The way you make a number positive depends on the sign of the number. If a number is positive you don't have to do anything to make it positive. If the number is negative, multiply it by -1 to make it positive.

Once you have isolated the absolute value, set its contents (what is inside the absolute value) equal to both the positive and the negative of whatever is on the other side of the equality. This is where you get to drop the absolute value symbol in the problem. You'll usually have two equations that you need to solve. Be sure to check your answers when you are done. Don't be surprised if both answers work. Often times there are several solutions to an algebraic equation. If that is the case, then you need to present all possible solutions.

Example 9: Solve for x: $|x| + 5 = 8$

Solution: First, isolate the absolute value using the tricks discussed earlier:

$$|x| + 5 = 8$$

$$|x| + 5 - 5 = 8 - 5$$

$$|x| = 3$$

Next, set the contents of the absolute value equal to the positive and negative values of whatever is on the other side of the equality:

$$|x| = 3$$

$$x = 3 \text{ or } x = -3$$

Finally, check your answers to be sure that they work: $|3| + 5 = 3 + 5 = 8$, so the first answer is okay. Since $|-3| + 5 = 3 + 5 = 8$, the second answer works as well. So the two values of x that satisfy the equality are $x = 3$ and $x = -3$.

Xtracts

When solving equations that contain an absolute value, you'll usually end up solving two equations.

Example 10: Solve for x: $|x + 3| = 7$

Solution: Since the absolute value is already isolated, you can jump to the next step and set the contents of the absolute value to +7 and –7:

$$x + 3 = 7 \text{ or } x + 3 = -7$$

Now solve each equality separately:

$x + 3 = 7$	$x + 3 = -7$
$x + 3 - 3 = 7 - 3$	$x + 3 - 3 = -7 - 3$
$x = 4$	$x = -10$

You have two answers: $x = 4$ and $x = -10$. Check them both to be sure that both are correct: $|4 + 3| = |7| = 7$, and $|-10 + 3| = |-7| = 7$, so both of them are okay. The values of x that satisfy the equality are $x = 4$ and $x = -10$.

Example 11: Solve for x: $2|x+5|+3=9$

Solution: Isolate the absolute value:

$$2|x+5|+3=9$$

$$2|x+5|+3-3=9-3$$

$$2|x+5|=6$$

$$\left(\frac{1}{2}\right)\cdot 2|x+5|=\frac{1}{2}\cdot 6$$

$$|x+5|=3$$

Next, set the contents of the absolute value equal to +3 and −3 and solve each equation:

$$x+5=3 \text{ or } x+5=-3;$$

$$x+5=3 \qquad\qquad x+5=-3$$

$$x+5-5=3-5 \qquad x+5-5=-3-5$$

$$x=-2 \qquad\qquad x=-8$$

The two answers are $x=-2$ and $x=-8$. Before you go on, check your work: $2|-2+5|+3=2|3|+3=2\cdot 3+3=6+3=9$, and $2|-8+5|+3=2|-3|+3=2\cdot 3+3=6+3=9$. Since both answers check out, the values of x that satisfy the equation are $x=-2$ and $x=-8$.

The Least You Need to Know

- The golden rule of algebra: Do unto one side of the equation as you do to the other.

- An equivalence relation is a relation that has the reflexive, symmetric, and transitive properties.

- To solve algebraic equations use the add-'em/subtract-'em and multiply-'em/divide-'em tricks to isolate the variable.

- To solve absolute-value problems, isolate the absolute value and then set its contents to the positive and negative values of whatever is on the other side of the equality.

Inequalities and Graphs

In This Chapter

- Describing inequalities using set notation
- Graphing inequalities on a number line
- Solving inequalities
- Mixing inequalities and absolute values

Mathematics is a reflection of the world in which we live. If our world was perfect, there wouldn't be inequalities, and you would be able to skip this chapter. But it's not, so you can't. In order to recognize inequalities, you have to be able to see them, and that means getting "graphic."

Properties of Inequalities

An inequality is a statement in which algebraic expressions are not equal. There are several symbols used to show inequality: > (greater than), ≥ (greater than or equal to), < (less than), and ≤ (less than or equal to). Inequalities have several special properties worth mentioning. Some of them will look familiar; equality and inequality have some of these properties in common.

As you saw earlier, inequalities have the transitive property: If $a > b$ and $b > c$, then $a > c$. You can substitute any of the four inequalities into that statement and it will hold true. Just make sure you use the same inequality throughout.

Inequalities are blessed with two addition properties. The nice thing about inequalities is that the balance has already been shifted; the not-so-nice thing about inequalities is that you are only allowed to either keep the imbalance or make things more unbalanced. No Robin Hood activities are allowed; you cannot give to the poor without giving as much (if not more) to the rich! The first addition property says that if $a > b$, then $a + c > b + c$. In other words, if you have an imbalance and you add the same thing to both sides of the statement, then you will not change the imbalance. The second addition property says that if $a > b$ and $c > d$, then $a + c > b + d$. This should make sense. For example, if I have $20 and you have $10 (I have more money than you do), and if someone gives me $30 and they give you $20, then I will still have more money than you do. The addition properties do not care whether c is positive or negative; you could make up subtraction properties that say the same thing.

Forewarning

With inequalities, the goal is not to right the wrong. If $a > b$ and $c > d$, you can't conclude anything about $a + d$ vs. $b + c$.

Things get a bit tricky when you multiply both sides of an inequality by a number. If the number you are multiplying by is positive, then what you expect to happen will happen: If $a > b$ and $c > 0$, then $ac > bc$. But if the number you are multiplying by is negative, things get a bit strange. If $a > b$ and $c < 0$, then $ac < bc$. Notice that the inequality flips over. Since division is the same as multiplying by the inverse, you could make up a division property that says the same thing. You would still have to distinguish between whether c was positive or negative, though.

The Real Deal

Properties of inequality:

If $a > b$ and $c > d$, then $a + c > b + d$.

If $a > b$ and $c > 0$, then $ac > bc$ and $\dfrac{a}{c} > \dfrac{b}{c}$.

If $a > b$ and $c < 0$, then $ac < bc$ and $\dfrac{a}{c} < \dfrac{b}{c}$.

Set Notation

To represent inequalities using sets, remember your set notation. Use curly brackets to enclose your description. Start with a variable name, then a divider (either | or :), and then the description. The description will include the appropriate inequality as well as whether you are working with real numbers, integers, rational numbers, etc …

Example 1: Write sets for the following:

a) All real numbers greater than 5 using set notation

b) All integers less than 7

c) All real numbers greater than or equal to 10

d) All rational numbers greater than 0 and less than 1

Solution:

a) $\{x : x > 5\}$

b) $\{x : x \text{ is an integer and } x < 7\}$

c) $\{x : x \geq 10\}$

d) $\{x : x \text{ is a rational number, } x > 0, \text{ and } x < 1\}$

The last example could be shortened. If a number is sandwiched between two other numbers, you are allowed to combine that information into one statement. Since 2 is between 0 and 5, you could write $0 < 2 < 5$, with the understanding that, if you needed to, you could split the statement into two inequalities: $0 < 2$ and $2 < 5$. So in example (d) you could have written $\{x : x$ is a rational number, $0 < x < 1\}$.

Solving Inequalities

Solving inequalities involves many of the same steps that solving equalities did. The only difference is that you have to pay attention to which side is bigger, and you have to be careful if you multiply or divide by a negative number.

Example 2: Solve the inequality $3x + 5 > 8$, and give your answer in set notation.

Solution: Just use your tricks:

$$3x + 5 > 8$$
$$3x + 5 - 5 > 8 - 5$$
$$3x > 3$$
$$\left(\frac{1}{3}\right)3x > \left(\frac{1}{3}\right)3$$
$$x > 1$$

In set notation, the solution is $\{x : x > 1\}$.

Example 3: Solve the inequality $4x - 6 < 2x + 10$, and give your answer in set notation.

Solution: Gather your x's on one side, the numbers on the other, make the coefficient in front of the x equal to 1 and you are done:

$$4x - 6 < 2x + 10$$
$$4x - 6 + 6 < 2x + 10 + 6$$
$$4x < 2x + 16$$
$$4x - 2x < 2x + 16 - 2x$$
$$2x < 16$$
$$\left(\frac{1}{2}\right)2x < \left(\frac{1}{2}\right)16$$
$$x < 8$$

In set notation, the solution is $\{x : x < 8\}$.

Example 4: Solve the inequality $2x + 10 \leq 5x + 16$, and give your answer in set notation.

Solution: Follow the same steps as in the previous examples:

$$2x + 10 \leq 5x + 16$$
$$2x + 10 - 10 \leq 5x + 16 - 10$$
$$2x \leq 5x + 6$$
$$2x - 5x \leq 5x + 6 - 5x$$
$$-3x \leq 6$$

Here comes the tricky part. When you multiply both sides of the inequality by $\left(-\frac{1}{3}\right)$, be sure to flip the inequality:

$$-3x \leq 6$$

$$\left(-\frac{1}{3}\right)(-3x)^3 \geq \left(-\frac{1}{3}\right)6$$

$$x \geq -2$$

In set notation, the solution is $\{x : x \geq -2\}$.

When you solve inequalities, the only time that you are allowed to change the inequality is when you multiply (or divide) by a negative number. Other than that, you work with the inequality that you were given throughout the problem.

Forewarning

If you multiply both sides of an inequality by a negative number, be sure to flip the inequality.

The Number Line

Inequalities are best visualized on a number line. The real number line is a pictorial representation of the real numbers; every number corresponds to a point. Some of the integers (not all of them!) are labeled and are evenly spaced on the number line. I would be remiss in my duties if I allowed you to draw number lines before negotiating the terms.

Every number line has an *origin*. It is the special point that represents 0. By convention, points to the left of the origin represent negative numbers and points to the right of the origin represent positive numbers.

A *ray* is half a line. A ray starts at a specified point and includes everything to either the left or the right of that point. The starting point may or may not be included in the ray. If the starting point is included in the ray, then the ray is *closed*; if the starting point is not included in the ray, then the ray is *open*.

An *interval* is a part of a line that has definite starting and stopping points. Intervals have two endpoints, and consist of everything between those two endpoints. The endpoints may or may not be included. If both endpoints are included, the interval is *closed*. If neither endpoint is included, then the interval is *open*. If only one of the endpoints is included, then the interval is *half open* (or half closed, depending on whether you are an optimist or a pessimist).

Got the Picture?

Number lines can help visualize equalities as well as inequalities. The point $x = a$ can be visualized as a filled-in dot, as shown in Figure 8.1. The point a is labeled. If a is positive, it will be drawn to the right of the origin; if a is negative, it will be drawn to the left of the origin.

a

Figure 8.1
The point x = a.

Inequalities lead to the creation of rays, as shown in Figure 8.2. If $x \geq a$, then the real numbers that satisfy this inequality include a and everything to the right of a. This is an example of a closed ray. If $x > a$, then the real numbers that satisfy this inequality include everything to the right of a, but not a itself. This is an example of an open ray. Unless otherwise stated, assume that x is a real number.

a)

a

b)

a

Figure 8.2

(a) The closed ray x $\geq$ a. *(b) The open ray* x $>$ a.

Rays that point to the left are created using the inequality *less than*, and are shown in Figure 8.3. If $x \leq a$, then the real numbers that satisfy this inequality include a and everything to the left of a. This is another example of a closed ray. If $x < a$, then the real numbers that satisfy this inequality include everything to the left of a, but not a itself. This is another example of an open ray.

a)

a

b)

a

Figure 8.3

(a) The closed ray x ≤ a. *(b) The open ray* x < a.

The rays shown in the previous figures look like part of the line. That's because you were allowing x to take on any real number that satisfied the specific inequality. If you restrict x so that it can only be an integer, things look a bit different. When visualizing the integers, keep in mind that integers are at least one unit away from each other. Real numbers can get really, really close to other real numbers. The set of integers that satisfies the inequality $x \geq a$ has to reflect the separation of the integers. The graph of the integers that satisfies the inequalities $x \geq a$, $x > a$, $x \leq a$, and $x < a$ are shown in Figure 8.4. Notice that instead of a solid line there are a bunch of closed circles, and possibly one open circle (if the equality option is not exercised).

a)

a

b)

a

c)

a

d)

a

Figure 8.4

(a) The closed ray x ≥ a, *where* x *is an integer. (b) The open ray* x > a, *where* x *is an integer. (c) The closed ray* x ≤ a, *where* x *is an integer. (d) The open ray* x < a, *where* x *is an integer.*

Example 5: Solve the inequality $4x + 4 > 2x + 8$ and graph your answer.

Solution: Isolate your variable using the same procedures in the last two examples:

$$4x + 4 > 2x + 8$$
$$4x + 4 - 4 > 2x + 8 - 4$$
$$4x > 2x + 4$$
$$4x - 2x > 2x + 4 - 2x$$
$$2x > 4$$
$$\left(\frac{1}{2}\right)2x > \left(\frac{1}{2}\right)4$$
$$x > 2$$

The graph of the inequality is shown in Figure 8.5.

Figure 8.5

The graph of the inequality x > 2.

Example 6: Graph the set $\{x : x < 2 \text{ or } x \geq 4\}$.

Solution: Graph both inequalities separately. Your answer includes all real numbers that satisfy one or the other (or both, if possible) of these inequalities, as is shown in Figure 8.6. Notice that you are looking at the union of the sets $\{x : x > 2\}$ and $\{x : x \geq 4\}$.

Figure 8.6

The graph of $\{x : x > 2 \text{ or } x \geq 4\}$.

Example 7: Graph the set $\{x : x \geq 1 \text{ or } x < 4\}$

Solution: Graph both inequalities separately. Your answer includes all real numbers that satisfy both of those inequalities, as is shown in Figure 8.7. Notice that you are looking at the intersection of the sets $\{x : x \geq -1\}$ and $\{x : x < 4\}$.

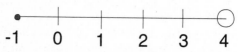

Figure 8.7

The graph of $\{x : x \geq -1 \text{ or } x < 4\}$.

Intervals

Points that lie in an interval satisfy two inequalities:
They are bigger than some minimum value (the left
endpoint), and they are less than some maximum
value (the right endpoint). The endpoints may or
may not be included. The interval described by the
inequalities $x \geq a$ and $x \leq b$, where x is real, is shown
in Figure 8.8(a). You can combine these inequalities
into one: $a \leq x \leq b$. Notice that a filled (closed) circle
is used for both endpoints; this is a closed interval.
Open intervals are drawn using unfilled (open) circles. Intervals that are half-open or half-closed use
one open circle and one closed circle. Figure 8.8
shows graphs of the four types of intervals.

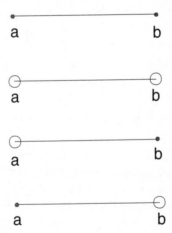

Figure 8.8

(a) The closed interval $a \leq x \leq b$. *(b) The open interval*
$a < x < b$. *(c) The half-open interval* $a < x \leq b$. *(d) The half-
open interval* $a \leq x < b$.

Notice that when you use a ≤ or ≥ symbol, your graph has a closed circle at the corresponding endpoint. When you use a > or < symbol, your graph will have an open circle at the corresponding endpoint.

Xtracts

Use a filled-in circle if the endpoint is included, and use an open circle if the endpoint is not included.

Interval Notation

You can use a shorthand notation to represent rays and intervals. Since the real numbers can get as large as they want, they are described as being "unbounded." Use the symbol ∞ to indicate that a ray goes on forever to the right, and −∞ to indicate that a ray goes on forever to the left.

The number line has no beginning and no end (it continues on forever in both directions), so it is written (−∞,∞). A ray only goes on forever in one direction, and stops at the endpoint. If the endpoint of a ray is included, use a square bracket; if the endpoint of the ray is not included, use parentheses. Figure 8.9 shows each type of ray and the corresponding inequality and interval notation.

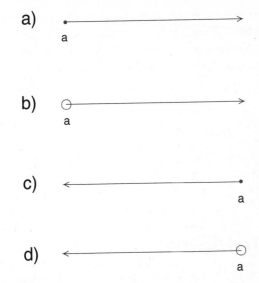

Figure 8.9

(a) The closed ray x ≥ a; interval notation: [a,∞). (b) The open ray x > a; interval notation: (a,∞). (c) The closed ray x ≤ a; interval notation: (–∞,a]. (d) The open ray x < a; interval notation: (–∞,a).

Now that you have seen how to represent rays, you can probably guess how to represent intervals. The smaller endpoint is written first and the larger endpoint is written second. A square bracket is used if the endpoint is included; otherwise, use parentheses. Figure 8.10 shows each type of interval, the corresponding inequality and interval notation.

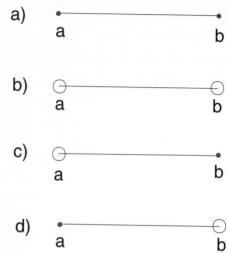

Figure 8.10

(a) The closed interval a ≤ x ≤ b; *interval notation:* [a,b]. *(b) The open interval* a < x < b; *interval notation:* (a,b). *(c) The half-open interval* a < x ≤ b; *interval notation:* (a,b]. *(d) The half-open interval* a ≤ x < b; *interval notation:* [a,b).

Example 8: Graph the interval $\{x : -3 < x \leq 2\}$.

Solution: The graph is shown in Figure 8.11.

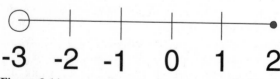

Figure 8.11

The graph of $\{x : -3 < x \leq 2\}$.

Example 9: Solve the inequality $2x - 7 \leq 3x + 2$, and give your answer in interval notation.

Solution: Use the same process you have been using, but give your answer in the requested form:

$$2x - 7 \leq 3x + 2$$

$$2x - 7 + 7 \leq 3x + 2 + 7$$

$$2x \leq 3x + 9$$

$$2x - 3x \leq 3x + 9 - 3x$$

$$-x \leq 9$$

$$(-1)(-x) \geq (-1)(9)$$

$$x \geq -9$$

The answer, in interval notation, is $[-9, \infty)$.

Inequalities and Absolutes

Solving inequalities that involve absolute values starts out the same way as solving equalities that involve absolute values; isolate the absolute value. You will then create two inequalities, one involving the positive and the other involving the negative of whatever is on the other side of the equality. There's just one twist: When you negate the stuff on the other side, you'll have to flip the inequality. After you have solved each inequality, you must think about how to combine the answers. Sometimes satisfying one or the other inequality is enough. Other times both inequalities have to be satisfied. I always recommend *thinking* about your answers rather than memorizing rules that may not make sense.

Xtracts _____

If an absolute value must be greater than a particular number, then the contents of the absolute value can be very large. In that case, your answer will usually be the *union* of two rays (one or the other inequality must be satisfied).

If an absolute value must be less than a particular number, then the contents of the absolute value can't get too big; the contents will typically be bounded (or finite) on both ends. In that case, your answer will usually be the *intersection* of two rays, (both inequalities must be satisfied).

But there are exceptions to these general observations.

Example 10: Solve the inequality $|x - 2| \leq 3$. Give your answer in set notation, interval notation, and graph your answer.

Solution: Since the absolute value is already isolated, create two inequalities. Since the absolute value must be less than or equal to 3, both inequalities will need to be satisfied:

$$x - 2 \leq 3 \qquad \text{and} \qquad x - 2 \geq -3$$
$$x - 2 + 2 \leq 3 + 2 \qquad\qquad x - 2 + 2 \geq -3 + 2$$
$$x \leq 5 \qquad\qquad\qquad x \geq -1$$

The answer, in set notation, is $\{x : x \geq -1 \text{ and } x \leq 5\}$. The answer, in interval notation, is $[-1, 5]$. The graph of the solution is shown in Figure 8.12.

Figure 8.12
The graph of $\{x : x \geq -1 \text{ and } x \leq 5\}$.

Example 11: Solve the inequality $|x + 1| > 3$. Give your answer in set notation, interval notation, and graph your answer.

Solution: Since the absolute value is already isolated, create two inequalities. Since the absolute value must be greater than 3, it is enough that either inequality is satisfied:

$$x + 1 > 3 \qquad \text{or} \qquad x + 1 < -3$$
$$x + 1 - 1 > 3 - 1 \qquad\qquad x + 1 - 1 < -3 - 1$$
$$x > 2 \qquad\qquad\qquad x < -4$$

The answer, in set notation, is $\{x : x < 4 \text{ and } x > 2\}$. The answer, in interval notation, is $(-\infty, -4) \cup (2, \infty)$. The graph of the solution is shown in Figure 8.13.

$$\longleftarrow \quad\!\!\!\!\!\overset{\text{-4}}{\circ}\quad\! \overset{\text{-3}}{|}\quad \overset{\text{-2}}{|}\quad \overset{\text{-1}}{|}\quad \overset{\text{0}}{|}\quad \overset{\text{1}}{|}\quad \overset{\text{2}}{\circ}\!\!\!\!\!\longrightarrow$$

Figure 8.13
The graph of $\{x : x < 4 \text{ and } x > 2\}$.

Example 12: Solve the inequality $3|x+1|-5 \le 1$. Give your answer in set notation, interval notation, and graph your answer.

Solution: Isolate the absolute value:

$$3|x+1|-5 \le 1$$
$$3|x+1|-5+5 \le 1+5$$
$$3|x+1| \le 6$$
$$\left(\frac{1}{3}\right) \cdot 3|x+1| \le \left(\frac{1}{3}\right) \cdot 6$$
$$|x+1| \le 2$$

Create the two inequalities. Since the absolute value must be less than or equal to 2, both inequalities will have to be satisfied:

$$x+1 \le 2 \qquad \text{or} \qquad x+1 \ge -2$$
$$x+1-1 \le 2-1 \qquad x+1-1 \ge -2-1$$
$$x \le 1 \qquad x \ge -3$$

The answer, in set notation, is $\{x : -3 \le x \le 1\}$. The answer, in interval notation, is $[-3,1]$. The graph of the solution is shown in Figure 8.14.

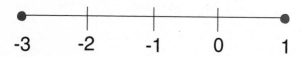

Figure 8.14
The graph of $\{x : -3 \le x \le 1\}$.

The Least You Need to Know

- Solve inequalities by using the add-'em/ subtract-'em trick or the multiply-'em/ divide-'em trick to isolate the variable.

- When multiplying an inequality by a negative number, be sure to flip the inequality.

- The graphs of inequalities are either rays or intervals. The rays and intervals can be open, half-open, or closed.

- When solving inequalities that also involve absolute values, you must flip the inequality when negating whatever is on the other side.

Relations, Functions, and Graphs

In This Chapter

- What it takes to be a relation
- The difference between a relation and a function
- Working with formulas and tables
- Graphing in the Cartesian coordinate system

The most abstract idea in algebra is the idea of a relation. You are probably already familiar with the ideas behind relations and functions, but you may not be used to looking at things through mathematical eyes.

Most people are quick to label mathematical relations as "too difficult to understand," but relations between people are far more complicated than relations between variables. In this chapter, I will help you understand relations between variables, but I'll defer the people problems to Dr. Phil.

Relations

If you have two sets, elements in one set can be connected (or related) with elements in the other set. Mathematicians use an *ordered pair* to make the connection (or relation) between elements explicit. An ordered pair consists of two elements, one from each set, separated by a comma and placed in parentheses. The object (1,15) is an example of an ordered pair, where 1 is an element in the first set and 15 is an element in the other set and the ordered pair (1,15) means that these two numbers (or elements of the set) are related to each other (maybe they are distant cousins).

A relation is a set of ordered pairs. The set {(1,1), (2,4), (3,9), (4,16)} is an example of a relation. In this case, the two sets that are being related are {1, 2, 3, 4} and {1, 4, 9, 16}. The first elements in all of the ordered pairs always come from the same set, and the last elements in the ordered pairs always come from the other set. The first set is called the *domain* of the relation and the second set is called the *range* of the relation.

The domain and range can have elements in common, and elements in the domain and the range can appear more than once in the relation.

The set {(1,1), (2,1), (1,4), (2,6)} is also a relation. The domain is the set {1, 2} and the range is {1, 4, 6}.

Example 1: Find the domain and range of the relation {(2,4), (6,8), (10,12), (14,16)}.

Solution: The domain is $\{2, 6, 10, 14\}$ and the range is $\{4, 8, 12, 16\}$.

Functions

A *function* is a relation that has a specific restriction: Each element in the domain is connected to one and only one element in the range. The relations $\{(1,1), (2,4), (3,9), (4,16)\}$ and $\{(1,1), (2,1), (3,1), (4,1)\}$ are functions, whereas the relation $\{(1,1), (2,2), (1,3), (2,4)\}$ is not. All functions are relations, but not all relations are functions.

Functions can be used to relate all kinds of things. A car's gas mileage can be thought of as a function of the speed that the car is driven (highway versus city driving). The temperature of New York City is a function of the time of the day and the day of the year. The closing value of the Dow Jones Industrial Average is a function of the number of sun spots generated during the day.

Forewarning

Do not mistake being a function of something with causality. Most mathematicians are not concerned with cause and effect. Functions can be created to relate events that are actually unrelated (like stock market prices and sun spots).

Functions have domains and ranges, since they are relations. Usually the domain is called X and the range is called Y. Generic elements in the domain are called x and generic elements in the range are called y. Functions also need names, and for lack of originality they are usually called $f(x)$ (which means that f acts on x) or just f.

Formulas

A function f can be interpreted as a set of instructions regarding what to do with elements in the domain, and they are typically defined by an equation or given by a formula. For example, the function that takes an element in the domain and first doubles it, and then adds 5, can be written using the formula $f(x) = 2x + 5$. Pay attention to the order of operations: Multiplication is done first, and then the addition. Think of f as a rule: Take whatever is in parentheses (in this case, it is x), double it, and then add 5. You can substitute specific values of x: $f(2) = 2(2) + 5 = 9$, $f(4) = 2(4) + 5 = 13$, and $f(W) = 2(W) + 5$. It doesn't matter what is inside the parentheses: f doubles it and adds 5.

Example 2: Describe the following functions in words, and evaluate the function for the specific value of x:

a) $f(x) = 4x + 7$; $x = 5$

b) $f(x) = 4(x + 7)$; $x = 6$

c) $f(x) = x^2$; $x = 4$

Solution:

a) This function takes an element in the domain and first multiplies it by 4 and then adds 7; $f(5) = 4(5) + 7 = 27$.

b) This function takes an element in the domain and first adds 7 and then multiplies the result by 4; $f(6) = 4(6 + 7) = 4(13) = 52$.

c) This function takes an element in the domain and squares it; $f(4) = 4^2 = 16$.

The Real Deal

While it is traditional to refer to functions as f or $f(x)$, any letter can be used for the function, and any letter can be used for the variable. The letters f, g, h, and y are the most commonly used letters for functions. The letter x is usually used for the variable, but the letter t is often used if the variable represents time.

Sometimes functions are thought of as magical boxes that transform the input (the variable) into output (the function value). They can be drawn schematically, as in Figure 9.1.

Figure 9.1

A function as a magical box.

Tables

Scientists often collect data and write functions in the form of a table. The tables can be arranged either vertically or horizontally. They include a heading to indicate the values of *x* and the values of *f(x)*. The function described in the following table does not have a specific formula associated with it, but certain function values can be read: $f(2) = 10$ and $f(8) = 20$.

x	*f(x)*
2	10
8	20

The function described in the following table also does not have a specific formula associated with it, but you can tell that $f(3) = 7$ and $f(6) = 11$.

x	3	6
$f(x)$	7	11

The tables I showed you only specified two data points, but tables can list thousands of data points or even more.

Example 3: Using the following table, find $f(5)$ and $f(10)$.

x	$f(x)$
0	10
5	17
10	23
15	5

Solution: Read the corresponding values off of the table: $f(5) = 17$ and $f(10) = 23$.

The Cartesian Coordinate System

Graphing a function involves graphing two things at a time: the variable value and the function value. One number line isn't enough; you will need two. Arrange the two number lines so that they are perpendicular to each other, as shown in Figure 9.2. This arrangement is called the Cartesian coordinate system. The horizontal number line is called the x-axis, and it will be used to keep track of the values of the variable. The vertical number line is called the y-axis, and it will be used to keep track of the values of the function. The point of intersection is called the origin.

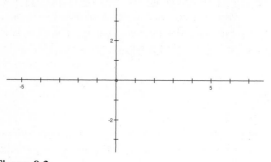

Figure 9.2

The Cartesian coordinate system.

The number line is a line and the Cartesian coordinate system is a plane. It is often referred to as the coordinate plane. The *x*-axis and the *y*-axis divide the plane into four regions or *quadrants*. These four quadrants are labeled using roman numerals (I, II, III, and IV) arranged in a counter-clockwise direction starting with the upper-right quadrant.

Every point in the plane can be thought of as an ordered pair *(a,b)*. The first coordinate is measured along the *x*-axis and the second coordinate is measured along the *y*-axis. The values of *a* and *b* indicate how far (and in which direction) the point is from the *y*- and *x*-axes respectively. The first element, *a*, describes how far the point is from the *y*-axis, where positive distances are measured to the right and negative distances are measured to the left. The second element, *b*, describes how far the point is from the *x*-axis, where positive distances are measured up and negative distances are measured down.

Xtracts

The coordinates of a point indicate the position of the point relative to the origin. The point (a,b) is located a units to the right (or left) of the origin (depending on its sign) and b units above (or below) the origin (depending on its sign).

Points located in Quadrant I have positive values for both of their x- and y-coordinates ($a > 0$ and $b > 0$). Points located in Quadrant II have negative values for their x-coordinate and positive values for their y-coordinates ($a < 0$ and $b > 0$). Points located in Quadrant III have negative values for both of their x- and y-coordinates ($a < 0$ and $b < 0$), and points located in Quadrant IV have positive values for their x-coordinates and negative values for their y-coordinates ($a > 0$ and $b < 0$). Points that are on the x-axis have their y-coordinate equal to 0 ($b = 0$), and points that are on the y-axis have their x-coordinate equal to 0 ($a = 0$).

Xtracts

The Quadrants and corresponding signs of the coordinates can be summarized: Quadrant I: $(+,+)$; Quadrant II: $(-,+)$; Quadrant III: $(-,-)$; Quadrant I: $(+,-)$.

Example 4: Graph points with coordinates (2,1) and (3,–4).

Solution: The solution is shown in Figure 9.3. The point with coordinates (2,1) is located 2 units to the right and 1 unit above the origin. The point with coordinates (3,–4) is located 3 units to the right and 4 units below the origin.

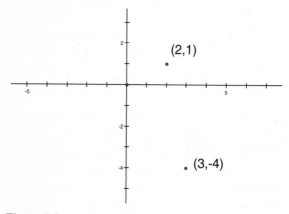

Figure 9.3

The graph of points with coordinates (2,1) and (3,–4).

Example 5: Give the coordinates of points P and Q shown in Figure 9.4.

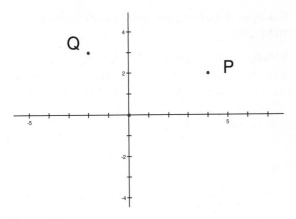

Figure 9.4
Points P and Q for Example 5.

Solution: The coordinates of point *P* are (4,2) and the coordinates for point *Q* are (–2,3)

Graphs

The point of describing the relationship between two concepts (such as gas mileage and speed) in terms of relations or functions is to gain an understanding of how those things are related. If you have a relation (or a function) specified by set notation (a set of ordered pairs), a formula (with a domain specified), or a table, you may not gain much insight into what you are trying to study. Since a picture is worth a thousand words, using the Cartesian coordinate system to generate a picture of your relation or function may help you analyze the situation.

If your function or relation is specified by a set of ordered pairs, then graph each ordered pair on the coordinate plane. If your function or relation is specified by a table, then view each row (or column) as an ordered pair and graph each ordered pair on the coordinate plane. If your function or relation is specified by an equation, then create a collection of ordered pairs by picking values in the domain and evaluating the function at those points. Graph these ordered pairs on the coordinate plane.

Example 6: Graph the following relations or functions using the Cartesian coordinate system. Determine which of them are relations and which are functions:

a) $\{(1,1), (2,1), (1,4), (2,6)\}$

b) $\{(1,1), (2,1), (3,1), (4,1)\}$

c)

x	0	3	6
$f(x)$	5	7	11

d) $f(x) = 3x - 5$, domain $\{1, 2, 3, 4\}$.

Solution: The graphs are shown in Figure 9.5.

a) There are two points with the same x-coordinate but different y-coordinates, so the set of ordered pairs is a relation.

b) Each ordered pair has a unique x-coordinate, so the set of ordered pairs is a function.

c) Each ordered pair has a unique x-coordinate, so the set of ordered pairs is a function.

d) Each ordered pair has a unique *x*-coordinate, so the set of ordered pairs is a function.

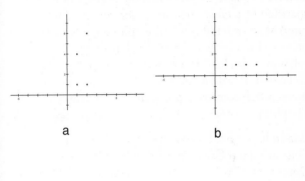

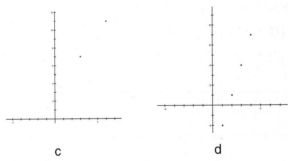

Figure 9.5

Graphs of: (a) $\{(1,1), (2,1), (1,4), (2,6)\}$ *(b)* $\{(1,1), (2,1), (3,1), (4,1)\}$

(c)

X	0	3	6
f(x)	5	7	11

(d) f(x) = 3x − 5, *domain* $\{1, 2, 3, 4\}$.

The Least You Need to Know

- Relations and functions are sets of ordered pairs.
- The domain is the set of all of the first elements in the ordered pairs. The range is the set of all of the second elements in the ordered pairs.
- Functions can be represented using set notation, formulas, tables, or graphs.
- The Cartesian coordinate system is a way to pictorially represent a function or relation.

Chapter 10

Linear Equations

In This Chapter

- The slope of a line
- The intercepts of a line
- Three forms of the equation of a line
- Three tricks for solving systems of equations

It is hard to find a place on this planet where lines or line segments cannot be found. Lines are fundamental building blocks, and linear equations are the language that an algebraist uses to talk about lines. Because lines come up again and again in mathematics, it behooves you to build a strong foundation now.

Lines

In Chapter 7 you learned how to solve linear equations with one variable. Now it's time to introduce equations with two variables. To keep things linear, each variable will only be raised to the first power. These equations will also involve real numbers.

Usually the variables involved in linear equations are *x* and *y*, but this is just customary and is not a requirement. Recall that when you draw the Cartesian coordinate system, the *x*-axis is the horizontal number line and the *y*-axis is the vertical number line. By convention, *x* is called the independent variable and *y* is called the dependent variable, because the value of *y* *depends* on the value of *x*.

Slopes

Lines can be horizontal, vertical, or inclined to the left or the right. The slope of a line is a measure of the steepness of its inclination. A horizontal line has a slope of 0 and a vertical line has an undefined slope. Inclined lines can have either a positive or a negative slope, depending on the direction of the incline.

Forewarning

"No slope" and "zero slope" are two different things. "No slope" means that there is no slope (the slope is not defined) and you have a vertical line. "Zero slope" means that the line is horizontal.

The slope of an inclined line can be found if you know the coordinates of any two points that lie on the line. If *(a,b)* and *(c,d)* are two points on the line, then the slope is calculated using the formula:

$$\text{slope} = \frac{d - b}{c - a}$$

This is equivalent to the equation:

$$\text{slope} = \frac{b-d}{a-c}$$

It doesn't matter which point is used first, but once you decide which one goes first you must stay consistent and use the same one first for both the numerator and the denominator.

The numerator of this ratio is the change in the y-coordinates of the points, and the denominator is the change in the x-coordinates of the points. The slope can be thought of as being the ratio of the "rise" (the change in y values) to the "run" (the change in the x values); the slope is rise over run. It is also the change in the dependent variable divided by the change in the independent variable (if letters other than x and y are used).

Example 1: Calculate the slope of the line passing through the following pairs of points:

a) (1,5) and (3,6)

b) (–1,6) and (2,4)

Solution: The slope is the change in y-coordinates divided by the change in the x-coordinates. It doesn't matter which point you use first as long as you are consistent.

a) $\text{slope} = \dfrac{6-5}{3-1} = \dfrac{1}{2}$

b) $\text{slope} = \dfrac{6-4}{-1-2} = \dfrac{2}{-3} = -\dfrac{2}{3}$

Intercepts

Lines can intersect the x-axis, the y-axis, or both. If a line does intersect the x- or y-axes, it will only do so in one place. The x-coordinate of the point where a line intersects the x-axis is called the *x-intercept*, and the y-coordinate of the point where a line intersects the y-axis is called the *y-intercept*. Horizontal lines (other than the x-axis) have no x-intercept, and vertical lines (other than the y-axis) have no y-intercept. Lines with an incline have both an x-intercept and a y-intercept.

Equations of Lines

There are many ways to write the equation of the line, but I will emphasize three: standard form, slope-intercept form, and point-slope form. While each form has its advantages and disadvantages, there's no law that says that you have to feel the same way about all three. I'm not ashamed to admit that I have a favorite, and you may have your favorite as well.

Point-Slope Form

The point-slope form of a line is $y - b = m(x - a)$, where a, b, and m are real numbers and x and y are the two variables. This form is usually an intermediate step in trying to write an equation in either standard form or slope-intercept form. In this equation, the line passes through the point (a,b) and has slope m.

Example 2: Find the point-slope equation of the lines with the following characteristics:

a) Passes through the point (5,2) and has slope 4

b) Passes through the points (2,–4) and (4,6)

Solution:

a) Since you are given the point and the slope, just substitute into the equation $y - b = m(x - a)$:

$$y - 2 = 4(x - 5)$$

b) You are given two points, but you need one point and the slope. You'll need to find the slope of the line first using the equation for the slope $\dfrac{d - b}{c - a}$:

$$\text{slope} = \frac{-4 - 6}{2 - 4} = \frac{-10}{-2} = 5$$

Now use the slope and either of the two points that were given, and substitute into the point-slope equation $y - b = m(x - a)$, where (a,b) is a point that the line passes through:

$$y - 6 = 5(x - 4)$$

This answer is equivalent to the equation $y - (-4) = 5(x - 2)$ (which you would get if you used the second point to get the equation).

Slope-Intercept Form

The slope-intercept form of a line is $y = mx + b$, where m and b are real numbers and x and y are the two variables. In this form, the slope and the y-intercept are immediately revealed (hence its name);

the slope is m and the y-intercept is b. If you are given the slope and the y-intercept, you can use the equation directly. Otherwise, you'll probably use the point-slope equation and simplify it; get y on one side of the equation, and everything else on the other side. Be sure to distribute the slope to get rid of the parentheses.

Example 3: Find the slope-intercept equation of the lines with the following characteristics:

a) Has slope 4 and y-intercept –6

b) Passes through (–3,2) and has slope 4

c) Passes through (1,–2) and (4,1)

Solution:

a) Since you are given the slope and the y-intercept, just substitute directly into the equation $y = mx + b$:

$$y = 4x + (-6) = 4x - 6.$$

b) You are given a point and the slope, so first use the point-slope formula and simplify to put the equation in the proper form:

$$y - 2 = 4(x - (-3))$$
$$y - 2 = 4(x + 3)$$
$$y - 2 = 4x + 12$$
$$y = 4x + 14$$

c) You need to find the slope first:

$$\text{slope} = \frac{-2-1}{1-4} = \frac{-3}{-3} = 1$$

Then use the point-slope formula (making use of either of the two points), and simplify. I'll use the second point:

$$y - 1 = 1(x - 4)$$

$$y - 1 = x - 4$$

$$y = x - 3$$

The slope is not immediately obvious because there is no coefficient in front of the x. Those of us "in the know" know that the slope is 1.

Standard Form

The standard form for a line is $Ax + By = C$, where A, B, and C are real numbers and x and y are the two variables. Quite often you will be given equations in standard form, and there are some teachers who only want you to write the equation of a line in standard form (I'm not one of them!). A and B can be used to calculate the slope of the line, C and B can be used to calculate the y-intercept of the line, and A and C can be used to find the x-intercept of the line.

Xtracts

If the equation of the line is in the standard form $Ax + By = C$, the slope is $-\dfrac{A}{B}$, the y-intercept is $\dfrac{C}{B}$, and the x-intercept is $\dfrac{C}{A}$.

When the equation of the line is given in standard form, information about the line is not readily available … if you don't remember the formulas given earlier then you'll have to do a little work. You will need to put the equation into slope-intercept form in order to determine the slope and y-intercept.

Example 4: Determine the slope and y-intercept of the line $3x + 2y = 7$.

Solution: You could identify A, B, and C and use the formulas for the slope $(-\frac{A}{B})$ and the y-intercept $(\frac{C}{B})$ and find the answer immediately: $A = 3$, $B = 2$, and $C = 7$, so the slope is $-\frac{3}{2}$ and the y-intercept is $\frac{7}{2}$.

Or, if you prefer to re-invent the equation, you could put the equation in slope-intercept form and read off the slope and the y-intercept directly. To put the equation in slope-intercept form, solve for y:

$$3x + 2y = 7$$

$$2y = -3x + 7$$

$$y = \frac{1}{2}\left(-3x + 7\right) = -\frac{3}{2}x + \frac{7}{2}$$

The slope is $-\frac{3}{2}$ and the y-intercept is $\frac{7}{2}$.

Example 5: Find the standard form of the equation of the line passing through $(3, -1)$ with slope 5.

Solution: Use the point-slope equation for a line and then write it in standard form:

$$y - (-1) = 5(x - 3)$$
$$y + 1 = 5x - 15$$
$$-5x + y = -16$$

It would have been all right to write your answer $5x - y = 16$ as well.

Xtracts

Given a line with slope m and y-intercept b, the slope-intercept equation of the line is $y = mx + b$ and the standard equation of the line is $mx - y = -b$.

Graphing Linear Equations

A line is uniquely determined by any two points. To graph a line in the coordinate plane, simply identify any two points on the line and graph them in the plane. Then take a straight edge and connect the points.

To find a point on the line, start by assuming a value for x and then find y. If all you need is any old point (or points), it's usually easiest to find the intercepts.

Example 6: Find the intercepts of the line given by the equation $2x + y = 4$.

Solution: To find the x-intercept, set $y = 0$ and find x:

$$2x + 0 = 4$$

$$2x = 4$$

$$x = 2$$

To find the y-intercept, set $x = 0$ and solve for y:

$$2 \cdot 0 + y = 4$$

$$0 + y = 4$$

$$y = 4$$

So the x-intercept is 2 and the y-intercept is 4.

Once you have two points, you can easily graph the line.

Forewarning

If the line passes through the origin, then the x-intercept and the y-intercept are both 0 and you will need to find another point in order to graph the line.

Example 7: Graph the line given by the equation $3x - 4y = 12$.

Solution: All you need is to find any two points, graph them on the coordinate plane, and connect them with a ruler or a straight edge. The easiest points to find are the intercepts:

First, find the *x*-intercept by setting *y* = 0:

$$3x - 4 \cdot 0 = 12$$
$$3x - 0 = 12$$
$$3x = 12$$
$$x = 4$$

That means that the line passes through the point (4,0). Next, find the *y*-intercept by setting *x* = 0:

$$3 \times 0 - 4y = 12$$
$$0 - 4y = 12$$
$$-4y = 12$$
$$y = -3$$

The line then passes through the point (0,–3). Now that you know two points (which happen to be the intercepts), graph them on the coordinate plane and connect them, as shown in Figure 10.1.

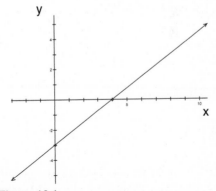

Figure 10.1
The graph of the line 3x – 4y = 12.

Example 8: Graph the line given by the equation $y = 2x$.

Solution: First, try to find the intercepts. Since the equation of this line is in slope-intercept form ($y = mx + b$, where b is the y-intercept) you can immediately read off the y-intercept; the y-intercept is 0. To find the x-intercept, set $y = 0$ and solve for x; in this case the x-intercept is also 0. So the x-intercept and the y-intercept lead to the same point (the origin). In this case you will need to find another point. Remember that you can find any two points. There is no rule that you have to use either of the intercepts. It's just that in my experience the intercepts are the easiest points to work with; you can't get much easier than letting one or the other of your variables equal 0! The second point can almost as easily be found by substituting $x = 1$ into the equation: $y = 2 \cdot 1 = 2$. So two points that the line passes through are (0,0) (the origin) and (1,2). Graph these two points on the coordinate plane and connect them, as shown in Figure 10.2.

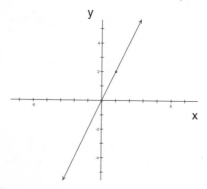

Figure 10.2

The graph of the line given by the equation $y = 2x$.

Systems of Linear Equations

Whenever you have two lines, only one of three things can happen: (a) one line lies on top of the other and they intersect at infinitely many places, (b) the two lines never intersect, or (c) the two lines intersect at one unique point.

If one line lies on top of the other, then the lines are *coincidental*; the two lines will have the same slope and the same *y*-intercept. If the two lines never intersect, they are *parallel*. Parallel lines are easy to recognize algebraically: They have the same slope but different *y*-intercepts. If the two lines have different slopes, they will intersect at one point; your goal will be to find that point.

The Real Deal _____

Given any two lines, exactly one of the following will be true:

- They intersect at infinitely many places.
- They intersect at exactly one point.
- They do not intersect.

Example 9: Classify the following systems of linear equations as coincidental (with infinitely many solutions), parallel (no point of intersection), or intersecting at a single point:

a) $\begin{cases} 2x + 3y = 5 \\ 4x + 6y = 10 \end{cases}$

b) $\begin{cases} x + 2y = 5 \\ 3x + 6y = 10 \end{cases}$

c) $\begin{cases} x + 2y = 5 \\ 2x - 6y = 0 \end{cases}$

Solution:

a) The slope of the first line is $-\frac{2}{3}$ and the y-intercept is $\frac{5}{3}$. The slope of the second line is $-\frac{4}{6} = -\frac{2}{3}$ and the y-intercept is $\frac{10}{6} = \frac{5}{3}$. Since the two lines have the same slope and the same y-intercept, the two lines are coincidental.

b) The slope of the first line is $-\frac{1}{2}$ and the y-intercept is $\frac{5}{2}$. The slope of the second line is $-\frac{3}{6} = -\frac{1}{2}$ and the y-intercept is $\frac{10}{6} = \frac{5}{3}$. Since the slopes are the same, but the y-intercepts are not, these lines are parallel.

c) The slope of the first line is $-\frac{1}{2}$ and the y-intercept is $\frac{5}{2}$. The slope of the second line is $-\frac{3}{6} = -\frac{1}{2}$ and the y-intercept is 0. Since the slopes are different, these two lines intersect at a single point.

A *system of linear equations* is just a collection of equations of lines, and the equations will be bracketed together. Usually you will have a system of two equations, and you will be asked to solve the system of equations. In solving the system of equations, you are looking for all points that satisfy both equations at the same time. That is why these problems are also known as solving *simultaneous equations*. There are several techniques that can be used to solve systems of equations. I will discuss three techniques: the substitution/elimination method, the addition/subtraction method, and the graphical method.

The Substitution/Elimination Method

This method involves solving for one of the variables in one of the equations and substituting into the other equation. This is the number-one method to use when one of the coefficients of one of the variables in one of the equations is 1.

Example 10: Solve the system of linear equations

$$\begin{cases} x + 2y = 3 \\ 2x + 3y = 4 \end{cases}$$

Solution: Since the coefficient of x in the first equation is 1, that will be the equation to use for our substitution. Isolate x in that equation:

$$x + 2y = 3$$
$$x = -2y + 3$$

Now use it to substitute into the second equation and solve for y:

$$2x + 3y = 4$$
$$2(-2y + 3) + 3y = 4$$
$$-4y + 6 + 3y = 4$$
$$-y + 6 = 4$$
$$-y = -2$$
$$y = 2$$

Now that you know what y is, you can use the first equation to solve for x:

$$x = -2y + 3$$
$$x = -2 \cdot 2 + 3$$
$$x = -1$$

Check to see that the point lies on both lines by substituting $x = -1$ and $y = 2$ into both original equations:

$$(-1) + 2 \cdot 2 = 3$$
$$2(-1) + 3 \cdot 2 = 4$$

The point that satisfies both linear equations is $(-1,2)$. Notice that your answer was a unique point. The lines in this system of equations intersect at a single point.

The Addition/Subtraction Method

This method is used when you can't use the substitution/elimination method. You will want to use the addition/subtraction method if none of the co-efficients of the variables are equal to 1. In this method you pick the variable that you will want to focus on (it doesn't matter which one you choose) and try to get the coefficients of that variable to be the same in both equations. The value that you want is the least-common multiple of the given coefficients. Then simply add (or subtract) the equations together (being careful to line up the equalities) to make that variable disappear.

Example 11: Solve the system of linear equations

$$\begin{cases} 3x + 2y = 1 \\ 2x + 3y = -1 \end{cases}$$

Solution: Since none of the coefficients are 1, the substitution/elimination method is not the preferred method. You could choose to focus on the x's or the y's; it really doesn't matter. I'll choose to focus on the x's. You want the x-coefficients to be the same value: the least-common multiple of 2 and 3. The smallest number that 2 and 3 both divide into evenly is 6. If you multiply the first equation by 2, then the x-coefficient of that equation will be 6. If you multiply the second equation by 3, then that equation's x-coefficient will also be 6. The system of equations will then be this:

$$\begin{cases} 6x + 4y = 2 \\ 6x + 9y = -3 \end{cases}$$

Now just subtract the bottom equation from the top equation:

$$6x + 4y = 2$$
$$- \ 6x + 9y = -3$$
$$\overline{0x - 5y = 5}$$

The resulting equation will lead to the value of y that satisfies both equations:

$$0x - 5y = 5$$
$$-5y = 5$$
$$y = -1.$$

Now that you know what y is, you can plug that value into either of the original equations to find x. I'll use the second equation:

$$2x + 3y = -1$$
$$2x + 3 \cdot (-1) = -1$$
$$2x - 3 = -1$$
$$2x = 2$$
$$x = 1.$$

The point that satisfies both equations is (1,–1). As always, be sure to check your answer in both original equations:

$$3 \cdot 1 + 2(-1) = 1$$
$$2 \cdot 1 + 3(-1) = -1.$$

The Graphical Method

The graphical method is just like it sounds. You graph each line and read off where they intersect (if they intersect). Unless your graphs are really accurate, this method only gives an estimate of where the point of intersection is. Whatever answer you come up with needs to be checked using both equations.

Example 12: Solve the system of linear equations

$$\begin{cases} x - y = 1 \\ x + y = 1 \end{cases}$$

Solution: Graph each line by finding two points (the intercepts) and determine where they intersect. The graph is shown in Figure 10.3, and from that graph it looks like the two lines intersect at the point (1,0).

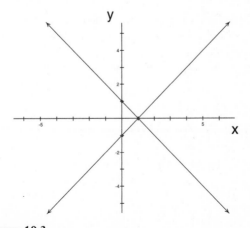

Figure 10.3

The intersection of the lines x − y = 1 *and* x + y = 1.

The point (1,0) satisfies both equations:

$$1 - 0 = 1$$
$$1 + 0 = 1$$

The estimates for x and y turned out to be exact solutions.

The Least You Need to Know

- The three forms of the equation of a line are point-slope, slope-intercept, and standard form.

- Horizontal lines have a slope of 0; vertical lines have no slope, or an undefined slope.

- The three methods for solving systems of equations are substitution/elimination, addition/subtraction, and the graphical method.

- Given any two lines, exactly one of the following will be true: They intersect at infinitely many places, they intersect at exactly one point, or they do not intersect.

Quadratic Equations

In This Chapter

- Factoring quadratic polynomials
- Solving quadratic equations
- Completing the square
- The quadratic formula

Quadratic equations are the most popular nonlinear polynomials. Graphs of quadratic equations are called parabolas. The shapes of satellite dishes and car headlamps are just two of many inventions that work because of the properties of parabolas. Because of the popularity of quadratic equations (and because I enjoy working with them so much), I have decided to devote an entire chapter to them.

Solving quadratic equations will usually involve factoring, so I might as well start there. You have already practiced some basic factoring techniques in Chapter 5. Factoring a quadratic polynomial can be thought of as just unFOILing the polynomial. The goal is to write the quadratic polynomial as a product of two linear polynomials, or *linear factors*.

The nice thing about factoring is that you can always check your work (just FOIL your answer to see if you are correct!). Factoring is a skill, and the more problems you work out, the more honed your skill will become. Of course, there are some insights that I'd like to share with you.

Factoring Quadratic Polynomials

In general, a quadratic polynomial has the form $ax^2 + bx + c$ where a, b, and c are real numbers. The leading term is ax^2, the middle term is bx, and the constant term is c. There was a time when factoring quadratic polynomials was a challenge even for mathematicians. Because mathematicians have a tenacity rivaled only by pit bulls, the secrets of quadratic polynomials are now well understood. Consider yourself among the chosen few to have that information revealed to you.

The quadratic polynomials that you will factor here will have integer coefficients. Before you dive head-first into the factoring waters, always check to see if you can factor out a monomial (or even just a constant, so that you are working with smaller numbers). If there are no monomials to pull out of each term, try the techniques illustrated in the following sections.

The Difference Between Two Squares

Many of the techniques for factoring stem from observations made while FOILing. For example, if you FOIL the product $(a + b)(a - b)$, you will end up with $(a + b)(a - b) = a^2 - a \cdot b + a \cdot b - b^2 = a^2 - b^2$, which is just the difference between two squares!

Working backward, then, you know that $a^2 - b^2 = (a + b)(a - b)$. Because of this observation, factoring the difference between two squares is almost effortless. All you have to do is take the square root of each term and then add them in one set of parentheses, and subtract them in the other; the two factors should be identical except for the + or – sign. Make sure that you pay attention to which square has the negative sign associated with it.

Example 1: Factor the following polynomials:

a) $x^2 - 1$

b) $y^2 - 9$

c) $z^2 - 12$

Solution:

a) $x^2 - 1 = (x + 1)(x - 1)$

b) $y^2 - 9 = (y + 3)(y - 3)$

c) $z^2 - 12 = \left(z + \sqrt{12}\right)\left(z - \sqrt{12}\right)$

Forewarning

Keep in mind that $a^2 + b^2$ cannot be factored. The fact that $a^2 - b^2$ (the difference between two squares) can be factored has to do with the negative sign in front of b^2.

Factoring When the Leading Coefficient is 1

To factor a quadratic polynomial whose leading coefficient is 1, you are looking for two linear factors, say $x + b$ and $x + d$, whose product is the quadratic polynomial given. The numbers b and d are referred to as the constant terms of the linear factors. If the leading coefficient is 1, you only have to pay attention to the signs and the coefficients of the x term and the constant term in the quadratic polynomial. The technique is best illustrated with an example.

Example 2: Factor the quadratic polynomial $x^2 + 6x + 5$.

Solution: You are looking for two linear factors, say $x + b$ and $x + d$, that, when multiplied, give $x^2 + 6x + 5$. In other words, $(x + b)(x + d) = x^2 + 6x + 5$. The product $b \cdot d$ must be 5 and their sum must be 6. The trick is to find two numbers that satisfy both conditions, so that when you FOIL your factors you get the original quadratic polynomial. Since 5 is a prime number, you know that the only pair of whole numbers whose product is 5 are 1 and 5. Fortunately for us, the sum of 1 and 5 is 6, and the only choice happens to be the right choice! So one of the constant terms has to be 5 and the other has to be 1. Putting the pieces of the puzzle together you have $(x + 5)(x + 1)$, and if you multiply these two factors together, you will get $x^2 + 6x + 5$.

There are certain clues to pay attention to when you are trying to factor a quadratic polynomial whose leading coefficient is 1. The sign of the constant term will tell you how to proceed. If the sign of the constant term is positive, then you are looking for factors of the constant term that *add up* to the co-efficient in front of x, and the constant terms in both factors will have the same sign as the coefficient of the x term in the quadratic polynomial. In other words, if the sign of the coefficient of the x term is positive, then the signs of both of the constants in the linear factors will also be positive. If the sign of the coefficient of the x term is negative, then the signs of both of the constants in the linear factors will also be negative. If the sign of the constant term is negative, then you are looking for factors of the constant term whose *difference* is the coefficient of the middle term. One of the constant terms will be positive and the other will be negative. To determine which constant gets which sign, look at the sign of the coefficient of the x term. Give the larger of the two factors of the constant term the same sign as the coefficient of the x term. The smaller of the two factors gets the opposite sign.

It may sound like a lot of rules, but once you start practicing, they will make more sense.

Example 3: Factor the following polynomials:

a) $x^2 + 2x - 8$

b) $x^2 - 2x - 15$

c) $x^2 - 5x + 6$

Solution:

a) To factor $x^2 + 2x - 8$, first look at the sign of the constant term. Since the constant term is negative, you are looking for two factors of 8 whose *difference* is 2. The factors of 8 are 8 and 1 (whose difference is 7 … not what you want), and 4 and 2 (whose difference is 2, which is what you want). Since the middle term is positive, so is the larger of the two factors of the constant term. The polynomial factors as $x^2 + 2x - 8 = (x + 4)(x - 2)$.

b) To factor $x^2 - 2x - 15$, look at the sign of the constant term. Since the constant term is negative, you are looking for two factors of 15 whose difference is 2. The factors of 15 are 15 and 1 (the difference is 14 … not what you want), and 5 and 3 (the difference is 2, which is what you want). Since the middle term is negative, so is the larger of the two factors of the constant term. The quadratic polynomial factors as $x^2 - 2x - 15 = (x - 5)(x + 3)$.

c) To factor $x^2 - 5x + 6$, look at the sign of the constant term. Since the constant term is positive, you are looking for two factors of 6 whose *sum* is 5. The factors of 6 are 6 and 1 (whose sum is 7 … not what you want), and 3 and 2 (whose sum is 5, which is what you want). Since the constant term of the quadratic polynomial is positive, the sign of the constant terms of the linear factors must be the same, and they must match the sign of the term that involves x in the quadratic polynomial. Since the term that involves x is negative, both of the constant terms in the linear factors will be negative. The quadratic polynomial factors as $x^2 - 5x + 6 = (x - 3)(x - 2)$.

Factoring When the Leading Coefficient is NOT 1

If the coefficient of the leading term is not 1, then you have to mix and match a bit more. You'll have to find the factors of the leading coefficient as well as the constant term in the quadratic polynomial and try to pair them up to give the correct answer.

You will probably have to try all combinations until you find the right one, at least at first. Believe me when I write that once you have worked "enough" problems, you'll start to recognize the factors and you'll turn into a lean, mean factoring machine. How many problems are enough? As many as it takes until you can recognize the factors fairly quickly! There is no "magic number."

Example 4: Factor the following polynomials:

a) $3x^2 + 10x + 8$

b) $4x^2 + 16x + 15$

Solution:

a) To factor $3x^2 + 10x + 8$, notice that the coefficient of the leading term is not 1, so you'll need to look at the factors of the leading coefficient and the constant term in the quadratic polynomial. Fortunately, in this case the leading coefficient is a prime number, so there's not as much mixing and matching necessary. The factors of the leading coefficient are 3 and 1, and the factors of 8 are either 8 and 1 or 4 and 2. Try all combinations, multiply them out (an opportunity to practice FOIL) and pick the right one. Only one will work:

$$(3x + 8)(x + 1) = 3x^2 + 11x + 8 \text{ (doesn't work)}$$

$$(3x + 1)(x + 8) = 3x^2 + 25x + 8 \text{ (doesn't work)}$$

$$(3x + 2)(x + 4) = 3x^2 + 14x + 8 \text{ (doesn't work)}$$

$$(3x + 4)(x + 2) = 3x^2 + 10x + 8 \text{ (it works!)}$$

Wouldn't you know it; it's always the last one you try!

b) To factor $4x^2 + 16x + 15$, notice that the coefficient of the leading term is not 1, so you will need to do a bit more work. In this case the leading coefficient is not a prime number, so there's even more mixing and matching necessary. But at least all of the signs are positive! The factors of the leading coefficient are either 4 and 1, or 2 and 2; the factors of 15 are either 15 and 1 or 5 and 3. Try all possible combinations:

$(4x + 15)(x + 1) = 4x^2 + 19x + 15$ $(4x + 3)(x + 5) = 4x^2 + 23x + 15$

$(4x + 1)(x + 15) = 4x^2 + 61x + 15$ $(2x + 15)(2x + 1) = 4x^2 + 32x + 15$

$(4x + 5)(x + 3) = 4x^2 + 17x + 15$ $(2x + 3)(2x + 5) = 4x^2 + 16x + 15$

So the correct way to factor $4x^2 + 16x + 15$ is $4x^2 + 16x + 15 = (2x + 3)(2x + 5)$.

Solving Quadratic Equations

A *quadratic equation* is an equation of the form $ax^2 + bx + c = 0$. Note the difference between a quadratic *polynomial* (or expression) $ax^2 + bx + c$ and a quadratic *equation*. With the equation, the expression *equals* something … 0. Look for the = sign; that is the key to knowing you are working with an equation. Solving quadratic equations involves finding the values of x that satisfy that equation. In other words, you are looking for the numbers to plug in for x in the quadratic polynomial to make it equal 0.

The key to solving a quadratic equation is to factor the quadratic expression into two linear factors.

That means that you take something that looks like $ax^2 + bx + c$ and turn it into something that looks like $(rx + s)(tx + u)$. Don't let the r, s, t, and u scare you; it's just that the symbols a, b, and c have been getting all of the attention!

Once you factor the quadratic expression, you will make use of one of the special properties of 0 mentioned in Chapter 1. Remember that if $a \cdot b = 0$, then either $a = 0$ or $b = 0$. Also, remember that 0 is the *only* real number with this property. It's a good trick to use, but it only works when the product of two factors equals 0. So by factoring the quadratic polynomial you will have two linear factors whose product is 0, and then because their product is 0, you know that one (or the other) factor must equal 0. Setting each linear factor equal to 0 gives you two new equations that are easier to solve. You will need to solve each of those linear equations for x, and you will get two answers to the question "What is x?" Both of those values of x are the solutions to the *original* quadratic equation.

Xtracts

Much of mathematics involves solving new problems using methods that have already been established. Solving quadratic equations is reduced to solving two linear equations.

As you solve more and more quadratic equations, you will begin to recognize some common threads or special tricks that you can use. These "factoring tricks" deserve special names. There are four standard methods for solving quadratic equations through factoring: factoring perfect squares, general factoring, factoring by completing the square, and the quadratic formula.

Perfect (and Not-So-Perfect) Squares

To solve the quadratic equation $x^2 = c$, rewrite it as $x^2 - c = 0$. You can easily factor the left side of the equation (it is the difference between two squares):

$$x^2 - c = 0$$
$$\left(x - \sqrt{c}\right)\left(x + \sqrt{c}\right) = 0$$

Now use the property of 0 by setting each of the two factors equal to 0 separately and solving for x:

$$\left(x - \sqrt{c}\right) = 0 \quad \text{or} \quad \left(x + \sqrt{c}\right) = 0$$

$$x = \sqrt{c} \quad \text{or} \quad x = -\sqrt{c}$$

So for the quadratic equation $x^2 = c$, either $x = \sqrt{c}$ or $x = -\sqrt{c}$. This can be written compactly as $x = \pm\sqrt{c}$.

Let's throw a wrench into the works; suppose you have a quadratic equation of the form $(x - a)^2 = c$. You can apply the same method just discussed to solve for $(x - a)$. If $(x - a)^2 = c$, then $\left(x - a\right) = \sqrt{c}$ or $\left(x - a\right) = -\sqrt{c}$.

Then you can solve for x by adding a to both sides of both equations:

$$x = a + \sqrt{c} \quad \text{or} \quad x = a - \sqrt{c}$$

These two equations can also be written more compactly: $x = a \pm \sqrt{c}$. If c is also a perfect square, then the solution can be simplified even further by combining a and $\sqrt{c}$, otherwise your answer will have to involve radicals.

Example 5: Solve the following quadratic equations:

a) $x^2 = 64$

b) $x^2 = 10$

c) $(x - 2)^2 = 9$

d) $(x + 1)^2 = 5$

Solution:

a) $x = \sqrt{64}$ or $x = -\sqrt{64}$, $x = \pm 8$

b) $x = \pm 10$

c) Since this problem involves perfect squares, it can be simplified: $(x - 2) = \pm\sqrt{9}$; $(x - 2) = 3$ or $(x - 2) = -3$; $x = 5$ or $x = -1$

d) Since this problem does not involve perfect squares, there's only so much you can do: $(x + 1) = \pm\sqrt{5}$; $x = -1 \pm \sqrt{5}$

General Factoring

To solve quadratic equations using the technique of general factoring, first factor the quadratic expression using the techniques discussed earlier in this chapter (see the section "Factoring Quadratic Polynomials"). Then set each linear factor equal to 0 and solve those equations.

Example 6: Solve the following quadratic equations by factoring:

a) $x^2 + 3x - 4 = 0$

b) $x^2 + 5x + 6 = 0$

c) $2x^2 + 5x + 2 = 0$

Solution: Factor each quadratic expression and then use the property of 0 to solve for x:

a) Factor the quadratic expression:

$$x^2 + 3x - 4 = (x + 4)(x - 1)$$

So the quadratic equation becomes $(x + 4)(x - 1) = 0$. Set each factor equal to 0 and solve for x:

$$(x + 4)(x - 1) = 0 \Rightarrow \text{either } (x + 4) = 0 \text{ or } (x - 1) = 0$$

$$(x + 4) = 0 \Rightarrow x = -4$$

$$(x - 1) = 0 \Rightarrow x = 1$$

So either $x = -4$ or $x = 1$. Check your answers in the original equation:

$$(-4)^2 + 3(-4) - 4 = 16 - 12 - 4 = 0$$

$$(1)^2 + 3(1) - 4 = 1 + 3 - 4 = 0$$

b) Factor the quadratic expression:

$$x^2 + 5x + 6 = (x + 2)(x + 3)$$

The quadratic equation becomes $(x + 2)(x + 3) = 0$.
Set each factor equal to 0 and solve for x:

$$(x + 2)(x + 3) = 0 \Rightarrow \text{either } (x + 2) = 0 \text{ or } (x + 3) = 0$$

$$(x + 2) = 0 \Rightarrow x = -2$$

$$(x + 3) = 0 \Rightarrow x = -3$$

So either $x = -2$ or $x = -3$. Check your answers:

$$(-2)^2 + 5(-2) + 6 = 4 - 10 + 6 = 0$$

$$(-3)^2 + 5(-3) + 6 = 9 - 15 + 6 = 0$$

c) Factor the quadratic expression:

$$2x^2 + 5x + 2 = (2x + 1)(x + 2)$$

The quadratic equation becomes $(2x + 1)(x + 2) = 0$. Set each factor equal to 0 and solve for x:

$$(2x + 1)(x + 2) = 0 \Rightarrow \text{either } (2x + 1) = 0 \text{ or } (x + 2) = 0$$

$$(2x + 1) = 0 \Rightarrow 2x = -1 \Rightarrow x = -\frac{1}{2}$$

$$(x + 2) = 0 \Rightarrow x = -2$$

So either $x = -\frac{1}{2}$ or $x = -2$. Check your answers:

$$2\left(-\frac{1}{2}\right)^2 + 5\left(-\frac{1}{2}\right) + 2 = 2 \cdot \frac{1}{4} - \frac{5}{2} + 2 = \frac{1}{2} - \frac{5}{2} + 2 = 0$$

$$2(-2)^2 + 5(-2) + 2 = 2 \cdot 4 - 10 + 2 = 8 - 10 + 2 = 0$$

Added Information

Factoring is an important skill. You must master factoring quadratic polynomials in order to solve quadratic equations correctly.

Completing the Square

Completing the square enables you to write a quadratic equation as a perfect (or not-so-perfect) square, so that you can use the methods discussed earlier. The goal is to rewrite the quadratic equation $ax^2 + bx + c = 0$ in such a way that the left side involves a perfect square. The first step involves rewriting the equation so that the terms that involve variables are on one side, and constants are on the other: $ax^2 + bx = -c$. At this point you must also make sure that the leading coefficient is 1; if $a \neq 1$, then divide both sides of the equation by a.

It's best to illustrate this technique with a specific equation: $x^2 - 6x + 8 = 0$. To complete the square, keep the terms that involve variables on one side of the equation and move the constant term over to the other side: $x^2 - 6x = -8$. The leading coefficient is already 1, so you are good to go onto the next step.

Take the coefficient in front of the x-term (–6), divide it in half, and then square it: $\left(\dfrac{-6}{2}\right)^2 = (-3)^2$. Take the resulting number (9) and add it to both sides of your equation: $x^2 - 6x + 9 = -8 + 9$. Because you are adding 9 to both sides of the equation, you are not upsetting the balance. The expression on the right is guaranteed to be a perfect square: $x^2 - 6x + 9 = (x - 3)^2$. If you don't believe me, FOIL it! So your equation turns into:

$$x^2 - 6x + 9 = -8 + 9$$

$$(x - 3)^2 = 1$$

Use the method of perfect (and not-so-perfect) squares to solve this equation:

$$(x - 3)^2 = 1$$

$$\text{either } (x - 3) = 1 \text{ or } (x - 3) = -1$$

$$x = 4 \text{ or } x = 2$$

Of course, you may have looked at the original equation and figured out how to factor it. If so, that's great! If you can identify the factors immediately, don't bother with this method. The strength of this method is that it can be used when you can't see the factors immediately, or the factors don't involve integers.

Example 7: Solve the following quadratic equations:

a) $x^2 + 4x - 2 = 0$

b) $2x^2 + 3x + 1 = 0$

Solution: The factors are probably not immediately obvious to you, so completing the square would be advisable.

a) Move the constant term over to the other side and make sure that the leading coefficient is 1:

$$x^2 + 4x - 2 = 0$$

$$x^2 + 4x = 2$$

Take the coefficient in front of the x-term (4), divide it in half, and then square it: $\left(\frac{4}{2}\right)^2 = (2)^2 = 4$. So add 4 to both sides of the equation, factor the left, and solve for x:

$$x^2 + 4x + 4 = 2 + 4$$

$$x^2 + 4x + 4 = 6$$

$$(x + 2)^2 = 6$$

$$(x + 2) = \pm\sqrt{6}$$

$$x = -2 \pm \sqrt{6}$$

The solutions are $x = -2 + \sqrt{6}$ or $x = -2 - \sqrt{6}$.

b) Move the constant term over to the other side and make sure that the leading coefficient is 1:

$$2x^2 + 3x + 1 = 0$$

$$2x^2 + 3x = -1$$

$$x^2 + \frac{3}{2}x = -\frac{1}{2}$$

Take the coefficient in front of the x-term ($\frac{3}{2}$), divide it in half, and then square it: $\left(\frac{3/2}{2}\right)^2 = \left(\frac{3}{4}\right)^2 = \frac{9}{16}$. So add $\frac{9}{16}$ to both sides of the equation, factor the left, and solve for x:

$$x^2 + \frac{3}{2}x + \frac{9}{16} = -\frac{1}{2} + \frac{9}{16}$$

$$x^2 + \frac{3}{2}x + \frac{9}{16} = -\frac{8}{16} + \frac{9}{16}$$

$$x^2 + \frac{3}{2}x + \frac{9}{16} = \frac{1}{16}$$

$$\left(x + \frac{3}{4}\right)^2 = \frac{1}{16}$$

$$\left(x + \frac{3}{4}\right) = \pm\sqrt{\frac{1}{16}}$$

$$\left(x + \frac{3}{4}\right) = \pm\frac{1}{4}$$

$$x = -\frac{3}{4} + \frac{1}{4} = -\frac{2}{4} = -\frac{1}{2} \quad \text{or} \quad x = -\frac{3}{4} - \frac{1}{4} = -\frac{4}{4} = -1$$

So the solutions are $x = -\frac{1}{2}$ or $x = -1$.

The Quadratic Formula

The quadratic formula is well known in algebraic circles. It may seem like a mystical formula handed down from the gods (which may explain why most people avert their eyes when it is presented!), but in reality, it is just a generalization of the process of completing the square. You can reason out the quadratic formula now that you understand the process of completing the square.

Here it is! The quadratic formula! Given the quadratic equation $ax^2 + bx + c = 0$ (with $a \neq 0$), the solutions are:

$$x = \frac{-b \pm \sqrt{b^2 - 4ac}}{2a}$$

Let's practice using this equation with the two examples worked in the previous section.

Example 8: Solve the following quadratic equations using the quadratic formula:

a) $x^2 + 4x - 2 = 0$

b) $2x^2 + 3x + 1 = 0$

Solution: The trick is to know what the values of a, b, and c are, and then put them into the formula correctly.

a) $a = 1$, $b = 4$, $c = -2$

$$x = \frac{-4 \pm \sqrt{4^2 - 4 \cdot 1(-2)}}{2 \cdot 1}$$

$$x = \frac{-4 \pm \sqrt{16 + 8}}{2}$$

$$x = \frac{-4 \pm \sqrt{24}}{2}$$

$$x = \frac{-4 \pm 2\sqrt{6}}{2}$$

$$x = \frac{\cancel{2}\left(-2 \pm \sqrt{6}\right)}{\cancel{2}}$$

$$x = -2 \pm \sqrt{6}$$

b) $a = 2, b = 3, c = 1$

$$x = \frac{-3 \pm \sqrt{3^2 - 4 \cdot 2 \cdot 1}}{2 \cdot 2}$$

$$x = \frac{-3 \pm \sqrt{9 - 8}}{4}$$

$$x = \frac{-3 \pm \sqrt{1}}{4}$$

$$x = \frac{-3 \pm 1}{4}$$

$$x = \frac{-3 + 1}{4} \quad \text{or} \quad x = \frac{-3 - 1}{4}$$

$$x = \frac{-2}{4} = -\frac{1}{2} \quad \text{or} \quad x = \frac{-4}{4} = -1$$

These answers look familiar!

Keep in mind that you have several methods to solve quadratic equations. If the x-term is missing in the quadratic equation, you will want to use the method of perfect squares. General factoring is quick *if* you see the factors. Of course, the only way to become adept at factoring is to work lots of problems. Completing the square lets you methodically solve the quadratic equation. After you have worked through several problems, the method should become second nature. Using the quadratic formula lets you solve every equation without having to reinvent the wheel each time, but you will have to reproduce the formula correctly every time. Put it to music and sing it on your way to school, write it out 1000 times, say it to yourself every night before you go to sleep ... do whatever it takes to commit it to long-term memory (you'll need it in Algebra II as well!). Be careful to substitute into the formula carefully, and take your time simplifying the expressions.

The Least You Need to Know

- To factor quadratic polynomials, you can use the method of perfect squares or analyzing the coefficients and unFOILing.

- The four techniques for solving quadratic equations are factoring perfect squares, general factoring, factoring by completing the square, and the quadratic formula.

- The quadratic formula: Given the quadratic equation $ax^2 + bx + c = 0$ (with $a \neq 0$), the solutions are $x = \dfrac{-b \pm \sqrt{b^2 - 4ac}}{2a}$.

Chapter 12

You Think You've Got (Word) Problems?

In This Chapter

- Percent problems
- The problem with integers
- Rate problems
- Money problems

While algebra is interesting in its own right, one of the reasons that it is an integral part of your mathematical education is because of how useful it is. The applications of algebra are illustrated in the form of word problems. These problems are as much fun to write as they are to solve! This chapter was originally 50 pages long, but my editor thought I went a bit overboard in my enthusiasm and had me trim it down. Mathematicians have tried to push the message that "algebra without word problems is like a day without sunshine" but for some reason the slogan hasn't caught on ....

In the algebra problems you have seen up to this point, I have been kind enough to give you the equations that you needed. With word problems, the subject of this chapter, you have to come up with the equations first, and then solve them.

How to Approach Word Problems

When you read a word problem, the first thing that you'll notice is that there are no variables or equations explicitly given. Your job is to create equations that fit the situation. Then you have to use the equations to answer whatever question is being asked. Sometimes the equations will be linear, other times they will be quadratic. For the quadratic equations, either factor or use the quadratic formula.

I recommend using a systematic approach when solving word problems. I'll discuss one approach that has worked well for me for years.

Identify the information you are given and what you are asked to find. Based on your understanding of the problem, choose variables to represent what you are given and what you are asked to find. Since you are going to create the equations, you can use any variables you want. I recommend writing down what your variables mean. You'll be surprised how easy it is to forget the goal of the problem. Writing down what your variables represent will help you get back on track if you get lost.

Read the word problem and try to translate the informa-
tion given in terms of your variables. Pay attention to the
units involved in each of your variables. Because the
English language is so rich, you will see many words
that represent addition, subtraction, multiplication,
and division. Your equation will involve addition if
you read words like sum, total, more, plus, or increase.
Your equation will involve subtraction if you read
words like difference, fewer, less than, reduced, or
decreased. Your equation will involve multiplication
if you read words like product, of, times, at, percent,
or twice. Your equation will involve division if you
read words like quotient, divided by, ratio, or half.

Solve the equation or perform the indicated calculation.
Use the properties of real numbers that you are
familiar with to solve the equation. You aren't allowed
to make up your own properties without a license.
After you are finished, check your answer in your
original equation.

Answer the question. Reread the problem to make
sure that you have answered the question. Think
about your answer and see if it makes sense.

Percent

Percents represent an amount out of 100. To convert
a number to a percentage, first write the number as
a decimal and then multiply by 100 (or move the
decimal two places to the right).

Example 1: Suppose that in a karate class 30% of the students have a brown belt. If there are 40 students in the class, how many of them have brown belts?

Solution: There are 40 students in the class, and you are asked to find the number of brown belts. Let x represent the number of brown belts (what you are asked to find). Set up a ratio:

$$\frac{\text{brown belts}}{\text{total students}} = \frac{30}{100}$$

Substitute the number of students and the number of brown belts into the ratio:

$$\frac{x}{40} = \frac{30}{100}$$

Now solve for x:

$$\frac{x}{40} = \frac{30}{100}$$

$$\cancel{40} \cdot \frac{x}{\cancel{40}} = 40 \cdot \frac{30}{100}$$

$$x = 12$$

Finally, answer the question. There are 12 students with brown belts.

Find-the-Integer Problems

"Find-the-integer" problems can make a long trip seem shorter (unless you don't like this sort of game, in which case the moments will drag on for what seems like an eternity!). In these problems, the sum (or difference) and product (or quotient) of two integers is given, and you must discover which integers are involved.

Example 2: The sum of two integers is 25 and one integer is four times as large as the other integer. Find the two integers.

Solution: Let x and y represent the two integers. Using variables, the first sentence can be translated into the equation $x + y = 25$. The second sentence can be translated into the equation $x = 4y$. Now you have a system of two equations and two unknowns:

$$\begin{cases} x + y = 25 \\ x = 4y \end{cases}$$

Since there is an equation with a variable that has a coefficient of 1, you can use the substitute/eliminate technique discussed in Chapter 10. Use the second equation to get rid of the variable x in the first equation by putting $4y$ everywhere you see an x:

$$4y + y = 25$$

$$5y = 25$$

$$y = 5$$

Now that you know y, use the second equation to find x:

$$x = 4y = 4 \cdot 5 = 20$$

The two numbers are 5 and 20. Check your answers: Is their sum 25 and is one number four times greater than the second? Yes.

Example 3: The difference between two integers is 15 and one integer is five more than twice the other. Find the two numbers.

Solution: Let x and y represent the two numbers. Translate the first sentence into an equation involving x and y: $x - y = 15$. Think about this equation: Since the difference between the two integers is positive, you are assuming $x > y$. This will be important when you write your second equation.

The second piece of information given is that one integer is five more than twice the other. Putting this information in terms of x and y can be done two ways: $x = 2y + 5$ or $y = 2x + 5$. One of these equations should be used in this problem and the other is an imposter. In order to determine which should be used, think about which one implies that $x > y$. In the first equation you have to add 5 to $2y$ in order to get x; that means that $x > y$.

If this is confusing, try plugging in a number for y, say $y = 1$, into both equations and finding the corresponding value of x. In the first equation, if $y = 1$, then $x = 2(1) + 5 = 7$ (and $7 > 5$). In the second equation, if $y = 1$, then $1 = 2x + 5$. Solving this equation for x you get

$$1 = 2x + 5$$

$$-4 = 2x$$

$$-2 = x$$

So if $y = 1$ then $x = -2$, and -2 is not greater than 1, so this second equation cannot be correct. It is the pretender to the throne.

You can also think about it in terms of money: if I have more money than you do, and someone doubles my money and then gives me $5 more, I'll have even more money than you. If $x > y$, the only way to restore balance is to give to the "have-nots" (y). Giving to the "haves" (x) will only create more of an imbalance.

No matter how you reason it out, the equation $x = 2y + 5$ is the equation of choice, and the system of equations needed to solve the problem are

$$\begin{cases} x - y = 15 \\ x = 2y + 5 \end{cases}$$

Since there are variables with coefficient equal to 1, the substitute/eliminate method will work well. In fact, since the second equation explicitly states how x and y are related, substitute the value of x from the second equation into the first equation:

$$(2y + 5) - y = 15$$

$$y + 5 = 15$$

$$y = 10$$

Use the value of y to solve for x:

$$x = 2y + 5 = 2 \cdot 10 + 5 = 25$$

So the two integers are 25 and 10. Check your answer: Is their difference 15? Yes. Is one integer equal to five more than twice the other? Yes. So your answer makes sense.

Example 4: The sum of two integers is 25 and their product is 84. Find the two integers.

Solution: Let x and y represent the two integers. The relationships between x and y can be written using the system of equations:

$$\begin{cases} x + y = 25 \\ x \cdot y = 84 \end{cases}$$

Use the first equation to eliminate one of the variables in the second equation. It doesn't matter which variable you eliminate:

$$y = 25 - x$$

$$x \cdot (25 - x) = 84$$

Distribute the x and then move everything to the same side to give a quadratic equation:

$$25x - x^2 = 84$$

$$-x^2 + 25x - 84 = 0$$

Multiply both sides of the equation by -1 to make the leading coefficient 1 (remember that):

$$x^2 - 25x + 84 = 0$$

You now have a quadratic equation that you need to solve. You can factor it (if you see the factors immediately) or use the quadratic formula. Since you are looking for two integers, the factors will be integers:

$$(x - 21)(x - 4) = 0$$

$$\text{either } (x - 21) = 0 \text{ or } (x - 4) = 0$$

$$x = 21 \text{ or } x = 4$$

There are two possible values of x, and each one will result in a different value for y:

$$x = 21 \Rightarrow y = 25 - 21 = 4$$

$$x = 4 \Rightarrow y = 25 - 4 = 21$$

No matter how you slice it, the two numbers are 4 and 21. Check your answer: Is their sum 25? Yes. Is their product 84? Yes.

Rate Problems

Rate problems always involve the same idea: Rate times time equals distance. When working word problems, I recommend that you avoid getting into the rut of always using x and y for your variables. Mix things up a little! If r stands for the rate, t stands for the time, and d represents distance, then the equation that relates all three variables is $r \cdot t = d$.

Example 5: The distance between Orlando, Florida, and Mishicot, Wisconsin, is 1370 miles. How long will it take Betty to drive from Orlando to Mishicot if she drove 65 miles per hour without stopping?

Solution: Use the equation $r \cdot t = d$, where r is the rate, t is the time, and d is the distance. If the rate is given in terms of miles per hour, then the distance should be in units of miles and the time should be in units of hours. Substitute the given information into the equation and solve for the variable that remains:

$$\frac{65\,\text{miles}}{\text{hours}} \cdot (t\,\text{hours}) = 1370\,\text{miles}$$

$$t = \frac{1370}{65} = 21\,\text{hours}$$

Answer the question: The time required is 21 hours. After doing this calculation, I'm sure Betty will check out airline rates!

Example 6: Jeanette is planning to drive 45 miles to a special bakery to get Denise some bread. Joe promised to have the lawn mowed before Jeanette returns. If it takes 15 minutes to have the bread wrapped and ready to ship, and Jeanette drives 40 miles per hour, how much time does Joe have to mow the lawn?

Solution: Use the equation $r \cdot t = d$, where r is the rate, t is the time, and d is the distance. If the rate is given in terms of miles per hour, then the distance should be in units of miles and the time should be in units of hours. There are three steps involved that require time: driving to the bakery, buying the bread, and driving back home. The time required to drive to the bakery is the same as the time required to return. You need to find the time it takes to drive to the bakery. Substitute the given information into the equation and solve for the variable that remains:

$$\frac{40\,\text{miles}}{\text{hour}} \cdot (t\,\text{hours}) = 45\,\text{miles}$$

$$t = \frac{45}{40} = \frac{9}{8} = 1.125\,\text{hours}$$

It takes 1.125 hours to drive each way, so the round trip takes 2.25 hours. To get the total time required for the bread purchase you must add the purchase time to the total driving time. The time that it takes to purchase the bread was given in minutes; you must first convert this time into hours before you add it to the driving time: 15 minutes is equivalent to .25 hours:

$$15 \text{ minutes} \cdot \frac{1 \text{ hour}}{60 \text{ minutes}} = .25$$

The total time involved is 2.25 hours (the total driving time) plus .25 hours (the purchase time), which is 2.5 hours. So Joe has 2.5 hours to mow the lawn.

Forewarning

When working with rate problems, make sure that your units for time are consistent: Either work in minutes throughout the problem or work in hours. Don't mix them up!

Money Problems

There's no guarantee that you will be able to solve all of your money problems after working these examples, but at least you'll have made a dent.

Example 7: Chelsea has some quarters and dimes in her pocket, worth $2.30. There are 14 coins total. To encourage you to develop your algebra skills, she makes a deal with you: If you can tell her how many quarters she has, then she will give you all of the money. How many quarters does she have?

Solution: You may be tempted to go back to using x and y for your variables; unfortunately they're on vacation, so you'll have to use some other letters. Let q represent the number of quarters and d represent the number of dimes. Since there are 14 coins total, you know that $q + d = 14$. Now you have to look at the value of the money. You have to decide if you are going to work with dollars or cents. As long as you are consistent, it doesn't matter. I'll set up the problem both ways, just so you can compare them.

First, let's work with cents. Each quarter is worth 25 cents and each dime is worth 10 cents, so the total amount of money Chelsea has based on her coinage is $25q + 10d$ cents. The total amount of money Chelsea has (as given in the problem statement) is 230 cents. So the equation is $25q + 10d = 230$, and the system of equation is:

$$\begin{cases} q + d = 14 \\ 25q + 10d = 230 \end{cases}$$

Suppose you had wanted to work with dollars. Then each quarter is worth .25 dollars, and each dime is worth .10 dollars. The total amount of money Chelsea has based on her coinage is $.25q + .10d$. The total amount of money Chelsea has (as given in the problem statement) is 2.30 dollars. So the equation is $.25q + .10d = 2.30$, and the system of equation is:

$$\begin{cases} q + d = 14 \\ .25q + .10d = 2.30 \end{cases}$$

Notice the differences between the two systems of equations. The first equation is the same in both cases, but the second equations are very similar ... there's a decimal that has to be included when working with dollars that doesn't have to be there when working with cents. Of course, the trade-off when working with cents is that the numbers are bigger (though they are whole numbers rather than decimals). It's six of one, half-dozen of the other, as they say in my favorite donut shop. It's time to solve the system of equations. Because it makes sense to me to work with cents, that's the system that I will use. Use the top equation to eliminate one of the variables from the bottom equation. Since the question asks for the number of quarters, I will use the top equation to get rid of the unwanted variable (dimes) in the bottom equation:

$$d = 14 - q$$

$$25q + 10(14 - q) = 230$$

$$25q + 140 - 10q = 230$$

$$15q + 140 = 230$$

$$15q = 90$$

$$\frac{1}{15} \cdot 15q = \frac{1}{15} \cdot 90$$

$$q = 6$$

While I am tempted to answer the question (Chelsea has 6 quarters), I can't check my work since I don't know how many dimes that means that Chelsea has. It's worth doing the extra work just to be able to check my work:

$$d = 14 - q = 14 - 6 = 8$$

If Chelsea has 6 quarters, then that means that she has 8 dimes. She has 14 coins, and the value of her coinage is $6 \cdot 25 + 8 \cdot 10 = 150 + 80 = 230$ cents. It's all good, and you can be confident in my answer: Chelsea has 6 quarters and now all of her money is mine, all mine!

Example 8: Aunt Kandi always gives her nieces and nephews money for their birthday. The amount she gives them depends on their age. If she gives them $5, plus 3 dollars for every year over 10, how much money will she give her nephew, Jeffrey, on his fifteenth birthday?

Solution: Let M represent the amount of money given, and let x denote Jeffrey's age; $(x - 10)$ then represents the number of years over 10. The amount of money given is

$$M = 5 + 3 \cdot (x - 10)$$

Since Jeffrey will be turning 15, $x = 15$; the amount of money he will receive on his birthday is

$$M = 5 + 3 \cdot (15 - 10) = 5 + 3 \cdot 5 = 5 + 15 = 20$$

So Jeffrey will get $20 from his Aunt Kandi.

Mixture Problems

Mixture problems require you to use conservation laws. Whatever you are mixing will be conserved and, if applicable, so will the money that the materials cost.

Example 9: The SunDeer Coffee House wants to mix two types of coffee, one that sells for $2.00 per pound and another that sells for $3.50 per pound, to get 30 pounds of coffee that costs $2.60 per pound. How many pounds of each type of coffee should be mixed?

Solution: Let C represent the number of pounds of coffee that costs $2.00 per pound, and P represent the number of pounds of coffee that costs $3.50 per pound. Since the goal is to make 30 pounds of the mixture, conservation of coffee gives the equation

$$C + P = 30$$

Now you have to conserve money. Each pound of the cheaper coffee costs $2.00; if the coffee house uses C pounds of the cheaper coffee, it will cost $2C$ dollars. Each pound of expensive coffee costs $3.50; if the coffee house uses P pounds of expensive coffee, it will cost $3.5P$. There are two ways to calculate the total cost of the mixture. Since the coffee house is making 30 pounds of coffee that costs $2.60 per pound, the total cost of the mixture is $30 \cdot 2.60 = 78$. Also, the total amount of money involved in making the mixture will be the sum of the costs involved using each type of coffee: $2C + 3.5P$. The rules of algebra require you to get the same answer regardless of which method is used, and that's how you get the second equation:

$$2C + 3.5P = 78$$

Now you have a system of two equations and two unknowns:

$$\begin{cases} C + P = 30 \\ 2C + 3.5P = 78 \end{cases}$$

Use the substitution/elimination method to solve for the two unknowns:

$$C = 30 - P$$
$$2(30 - P) + 3.5P = 78$$
$$0 - 2P + 3.5P = 78$$
$$60 + 1.5P = 78$$
$$1.5P = 18$$
$$P = \frac{18}{1.5} = 12$$

If you use 12 pounds of the expensive coffee, you will use 18 pounds of the cheaper coffee ($C = 30 - P = 30 - 12 = 18$).

Now that you've seen the variety of problems you can solve using your newly developed algebra skills, I'm sure you'll start looking for algebra problems in your day-to-day life. Fortunately, you won't have to look very far. You can write equations to help you figure out which cell-phone plan fits you best, or you can calculate how many hours you will have to work in order to buy that car you've had your eye on. Your newly developed algebra skills will help you calculate whatever you need to know, and if you're skills ever get rusty, you can always turn to this book for a quick review!

The Least You Need to Know

- When working out word problems, translate the given information into an equation or a system of equations that you can then solve.
- Rate times time equals distance. Pay attention to the units for time and distance.
- Setting up mixture problems involves the conservation of the substance as well as the conservation of money spent to make the mixture.

Glossary

absolute value of a   The magnitude of a.

additive identity   The unique number that you can add to any other number and have no effect.

additive inverse of a   The unique number that you add to a in order to get 0.

associative property of addition
$a + (b + c) = (a + b) + c$.

associative property of multiplication
$a \cdot (b \cdot c) = (a \cdot b) \cdot c$.

cardinality of a set   The number of elements in the set.

coefficient   The number in front of the variable.

commutative property of addition $a + b = b + a$

commutative property of multiplication
$a \cdot b = b \cdot a$.

complement   If A is a subset of U, then its complement, denoted $\bar{A}$ or A^c, is the set of all of the things in U that are not in A.

complex fraction A fraction where the numerator, denominator, or both contain a fraction.

composite number A number that can be evenly divided by numbers other than 1 and itself.

degree of a monomial The sum of the exponents of all of the variables involved in the monomial.

degree of a polynomial The degree of the largest monomial that is contained within the polynomial.

distributive property $a \cdot (b + c) = a \cdot b + a \cdot c$.

equation A mathematical statement that says two mathematical expressions are equal.

equivalence relation A relation that has the reflexive, symmetric, and transitive properties.

even number A natural number that is divisible by 2.

expression A statement that combines numbers and variables in a meaningful way using mathematical operations.

function A relation where each element in the domain is connected to one and only one element in the range.

inequality A statement in which algebraic expressions are not equal.

integers The natural numbers (or the positive integers), their negatives, or opposites, (or the negative integers), and 0. They can be written as … $-3, -2, -1, 0, 1, 2, 3$ ….

intersection of two sets   The set of all of the elements that are in both sets.

interval   Part of a line that has definite starting and stopping points.

irrational number   A number that cannot be written as a ratio of two integers.

monomial   An algebraic expression that consists of one term. The term can have many parts, but all parts are multiplied (or divided) together.

multiplicative identity   The only number that you can multiply any other number by and have no effect.

multiplicative inverse of a   The unique real number that you multiply a by in order to get 1.

negative integers   The numbers $-1, -2, -3$ ….

odd numbers   Natural numbers that are not divisible by 2.

polynomial   An expression that involves the sum of two or more monomials.

prime numbers   Natural numbers that can be divided evenly only by 1 and themselves.

principal root   The positive root.

quadratic equation   An equation of the form $ax^2 + bx + c = 0$.

quadratic polynomial   A polynomial in the form $ax^2 + bx + c$.

radical   The root sign $\sqrt{}$.

radicand The expression under the radical sign.

rational expressions Fractions that involve polynomials.

rational number A number that can be written as the ratio of two integers.

ray Half a line.

real number Any rational and irrational number.

real number line A pictorial representation of the real numbers.

reciprocal The multiplicative inverse.

relation A set of ordered pairs.

relatively prime numbers Two numbers that share no common divisors.

subset The set A is a subset of U if all of the elements in A are elements of U.

system of linear equations A collection of equations of lines.

transitive property of equality If $a = b$ and $b = c$, then $a = c$.

trichotomy property Given two real numbers a and b, exactly one of the following is true: $a < b$, $a = b$, or $a > b$.

union of two sets The set of all elements that are in either set (or in both).

Venn diagram A visual way to show the relationship between two or more sets.

whole numbers The numbers 0, 1, 2, 3 …

Properties and Formulas

In this book, you have been introduced to quite a few symbols, rules, and equations. I have collected them together here for an easy reference. Feel free to refer to this section if you get stuck on a problem or you just need a quick review.

Symbols Used

Symbol	Meaning
+	addition
–	subtraction
· or ( )( )	multiplication
/ or ÷	divided by
=	is equal to
≠	is not equal to
>	is greater than
≥	is greater than or equal to
<	is less than
≤	is less than or equal to

Properties of Real Numbers

$a - b = a + (-b)$

$a - (-b) = a + b$

$-a = -1 \cdot a$

$a \cdot (b - c) = a \cdot b - a \cdot c$

$-(a + b) = -a - b$

$-(a - b) = -a + b$

$(-a)(-b) = a \cdot b$

$0 \cdot a = 0$

$\dfrac{a}{1} = a$

$\dfrac{0}{a} = 0$ (a≠0)

$\dfrac{a}{a} = 1$ (a≠0)

$a\left(\dfrac{b}{a}\right) = b$ (a≠0)

Properties of Inequality

If $a > b$ and $c > d$, then $a + c > b + d$.

If $a > b$ and $c > 0$, then $ac > bc$ and $\dfrac{a}{c} > \dfrac{b}{c}$.

If $a > b$ and $c < 0$, then $ac < bc$ and $\dfrac{a}{c} < \dfrac{b}{c}$.

Rules for Exponents

$$a^n \times a^m = a^{n+m}$$

$$\frac{a^n}{a^m} = a^{n-m}; \quad \frac{a^n}{a^m} = \frac{1}{a^{m-n}}$$

$$(a^n)^m = a^{n \times m}$$

$$\frac{1}{a^{-n}} = a^n$$

$$a^0 = 1 \text{ if } a \neq 0$$

$$(a \cdot b)^n = a^n \cdot b^n$$

$$\left(\frac{a}{b}\right)^n = \frac{a^n}{b^n}$$

Rules for Roots

$$\sqrt[n]{a} = a^{\frac{1}{n}}$$

$$a^{\frac{n}{m}} = \left(a^{\frac{1}{m}}\right)^n = \left(\sqrt[m]{a}\right)^n$$

$$a^{\frac{n}{m}} = \left(a^n\right)^{\frac{1}{m}} = \sqrt[m]{a^n}$$

$$\sqrt[n]{a} \cdot \sqrt[m]{a} = \sqrt[mn]{a^{m+n}}$$

$$\frac{\sqrt[n]{a}}{\sqrt[m]{a}} = \sqrt[mn]{a^{m-n}}$$

$$\sqrt[m]{\sqrt[n]{a}} = \sqrt[mn]{a}$$

$$\sqrt[n]{\frac{a}{b}} = \frac{\sqrt[n]{a}}{\sqrt[n]{b}}$$

$$\sqrt[n]{a} \cdot \sqrt[m]{b} = \sqrt[mn]{a^m b^n}$$

Square of Sums

$(a + b)^2 = a^2 + 2a \cdot b + b^2$

Square of Differences

$(a - b)^2 = a^2 - 2(a \cdot b) + b^2$

Slope of a Line

If (a,b) and (c,d) are two points on the line, then the slope is calculated using this formula: slope $= \dfrac{d - b}{c - a}$.

Equations of Lines

point-slope form: $y - b = m(x - a)$

slope-intercept form: $y = mx + b$

standard form: $Ax + By = C$

If the equation of the line is in standard form, the slope is $-\dfrac{A}{B}$, the y-intercept is $\dfrac{C}{B}$, and the x-intercept is $\dfrac{C}{A}$.

quadratic formula: Given the quadratic equation $ax^2 + bx + c = 0$ (with $a \neq 0$), the solutions are $x = \dfrac{-b \pm \sqrt{b^2 - 4ac}}{2a}$.

Index

S

T